M000278644

McGraw-Hill

My Math

This is your very own math book! You can write in it, draw, circle, and color as you explore the exciting world of math.

Let's get started. Grab a crayon and draw a picture that shows what math means to you.

Have fun!

This is your space to draw.

McGraw Hill Education

MHEonline.com

STEM McGraw-Hill is committed to providing instructional materials in Science, Technology, Engineering, and Mathematics (STEM) that give all students a solid foundation, one that prepares them for college and careers in the 21st century.

Send all inquiries to:
McGraw-Hill Education
8787 Orion Place
Columbus, OH 43240

ISBN: 978-0-07-668780-0 (*Volume 1*)
MHID: 0-07-668780-5

Printed in the United States of America.

7 8 9 10 11 12 LMN 23 22 21 20

Meet The Artists!

Wilmer Cortez Cabrera

Numbers in my Life When we heard that I was a winner, my class friends hugged me so much that I fell on the floor. I feel like I am a star. *Volume 1*

Samantha Garza

I Add and Subtract I like to read, dance and play. Making this art work was fun. *Volume 2*

Other Finalists

K. Jock's and M. Kennedy's Class*
Time and Money is Math

Carly Gordon
Math and Art go great together!

Manuel Otero
Line Math

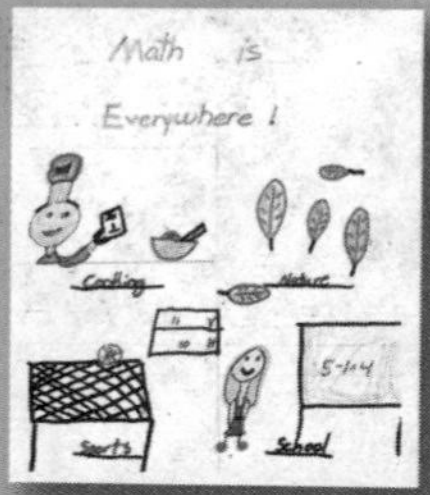

Katy Rupnow
Math is Everywhere!

Ma Myat Thiri Kyaw
Math Swamp

Jahni Williams
All About Numbers

Nora Carter's Class
Math for Life

Brittany Schweitzer
Serving up Math

Lillian Gaggin
Wristwatch

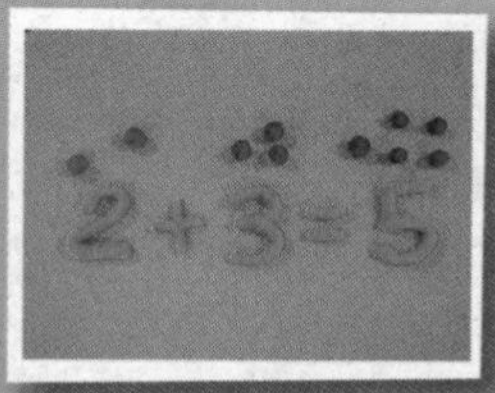

Kristie Mendez's Class*
Add Up the Dough

Find out more about the winners and other finalists at www.MHEonline.com.

We wish to congratulate all of the entries in the 2011 *McGraw-Hill My Math* "What Math Means To Me" cover art contest. With over 2,400 entries and more than 20,000 community votes cast, the names mentioned above represent the two winners and ten finalists for this grade.

** Please visit mhmymath.com for a complete list of students who contributed to this artwork.*

GO digital

it's all at
connectED.mcgraw-hill.com

Go to the Student Center for your eBook, Resources, Homework, and Messages.

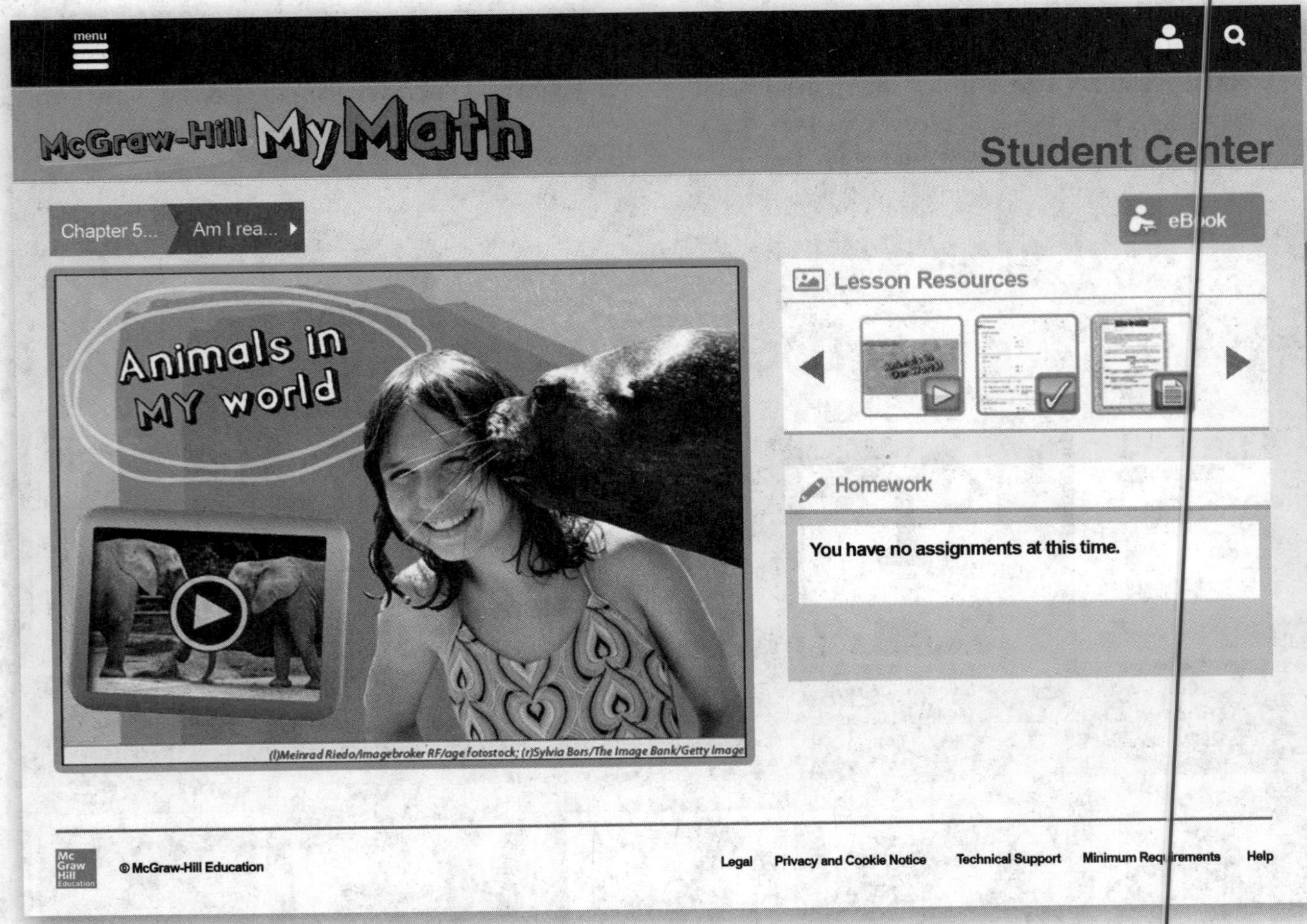

Write your Username

Password

Get your resources online to help you in class and at home.

Vocab

Find activities for building vocabulary.

Watch

Watch animations of key concepts.

Tools

Explore concepts with virtual manipulatives.

eHelp

Get targeted homework help.

Games

Reinforce with games and apps.

Tutor

See a teacher illustrate examples and problems.

GO mobile

Scan this QR code with your smart phone* or visit mheonline.com/stem_apps.

*May require quick response code reader app.

Available on the App Store

Contents in Brief

Organized by Domain

Processes & Practices Woven Throughout

Chapter 1

Addition Concepts

ESSENTIAL QUESTION
How do you add numbers?

Getting Started

Camping is a hoot!

Lessons and Homework

Wrap Up

Let's explore more online!

connectED.mcgraw-hill.com

Chapter 2 Subtraction Concepts

ESSENTIAL QUESTION
How do you subtract numbers?

Getting Started

Lessons and Homework

Wrap Up

connectED.mcgraw-hill.com

Your safari adventure starts online!

Chapter 3 Addition Strategies to 20

ESSENTIAL QUESTION
How do I use strategies to add numbers?

Getting Started

Let's roll into the big city!

Lessons and Homework

Wrap Up

Look for this!
Watch
Click online and you can watch videos that will help you learn the lessons.

connectED.mcgraw-hill.com

Chapter 4 Subtraction Strategies to 20

ESSENTIAL QUESTION
What strategies can I use to subtract?

Getting Started

Lessons and Homework

I love the beach!

Wrap Up

Look for this!
Click online and you can find activities to help build your vocabulary.
Vocab
abc

connectED.mcgraw-hill.com

Chapter 5 Place Value

ESSENTIAL QUESTION
How can I use place value?

Getting Started

Lessons and Homework

We're going to the toy store!

Wrap Up

There are fun games online!

connectED.mcgraw-hill.com

Chapter 6 Two-Digit Addition and Subtraction

ESSENTIAL QUESTION
How can I add and subtract two-digit numbers?

Getting Started

Lessons and Homework

Wrap Up

You can find fun activities online!

connectED.mcgraw-hill.com

Chapter 7 Organize and Use Graphs

ESSENTIAL QUESTION
How do I make and read graphs?

Getting Started

Let's get active!

Lessons and Homework

Wrap Up

Look for this!

Tools

Click online and you can find tools that will help you explore concepts.

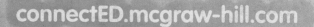

connectED.mcgraw-hill.com

Chapter 8

Measurement and Time

ESSENTIAL QUESTION
How do I determine length and time?

Getting Started

Look! I'm a watch dog!

Lessons and Homework

Wrap Up

My classroom is fun!

connectED.mcgraw-hill.com

Chapter 9 Two-Dimensional Shapes and Equal Shares

ESSENTIAL QUESTION
How can I recognize two-dimensional shapes and equal shares?

Getting Started

Lessons and Homework

We're going to the farm!

Wrap Up

Look for this!
Click online and you can check your progress.

Check

connectED.mcgraw-hill.com

Chapter 10 Three-Dimensional Shapes

ESSENTIAL QUESTION
How can I identify three-dimensional shapes?

Getting Started

Lessons and Homework

Wrap Up

Look for this!
Click online and you can get more help while doing your homework.
eHelp

connectED.mcgraw-hill.com

Chapter

1 Addition Concepts

ESSENTIAL QUESTION

How do you add numbers?

We're Going Outdoors!

Watch a video!

Watch

Name ....

Chapter 1 Project

Treasure Hunt

1. Work with your group to solve the first problem. Check your answer at the location given to you by your teacher. Follow the clue to the next location.
2. Solve the second problem. Check your answer. Follow the clue to the next location.
3. Repeat the steps until your group has solved all four problems correctly.

1. Andrew has 4 cats. Zada has 0 cats. How many cats do they have altogether?

______ cats

2. Kylie has 3 hamsters. Nate has the same number of hamsters. How many hamsters do they have?

______ hamsters

3. There are 10 cows and pigs on a farm. Write a way that could show the number of cows and pigs on the farm.

______ + ______ = 10 cows and pigs

4. There are 5 fish in a tank. 4 snails are also in the tank. How many animals are in the tank?

______ animals

Name ....

Am I Ready?

Write how many.

1. 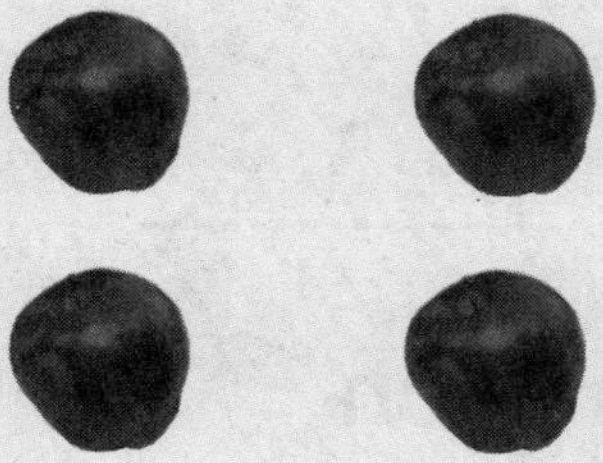 ______

2.  ______

Draw circles to show each number.

3. 5

4. 7

Write how many there are in all.

5. 

______ birds

How Did I Do?

Shade the boxes to show the problems you answered correctly.

1	2	3	4	5

Name

My Math Words

Vocab

Review Vocabulary

in all   same

Trace the words. Then draw a picture in each box to show what each word means.

Word Set	My Example
in all	
same	

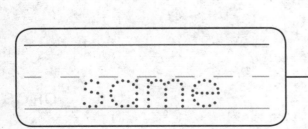

My Vocabulary Cards

Vocab abc

Processes & Practices

Lesson 1-2

add

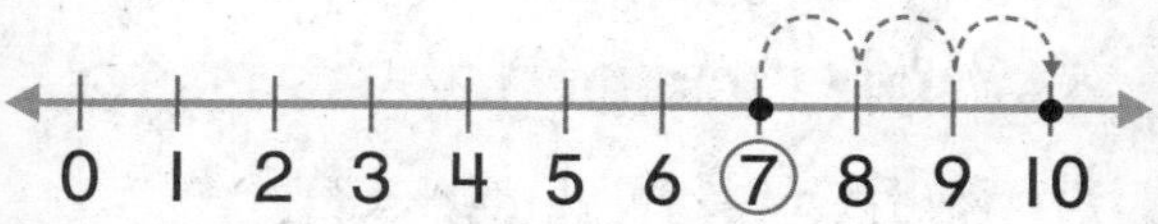

7 + 3 = 10

Lesson 1-3

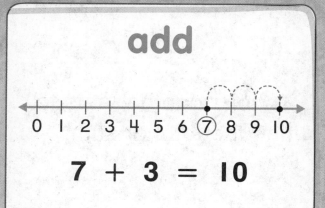

addition number sentence

4 + 4 = 8

9 = 6 + 3

Lesson 1-3

equals (=)

2 + 1 = 3

Lesson 1-13

false

3 + 1 = 5 is false

Lesson 1-2

part

Part	Part
5	2
Whole	
7	

part

Lesson 1-3

plus (+)

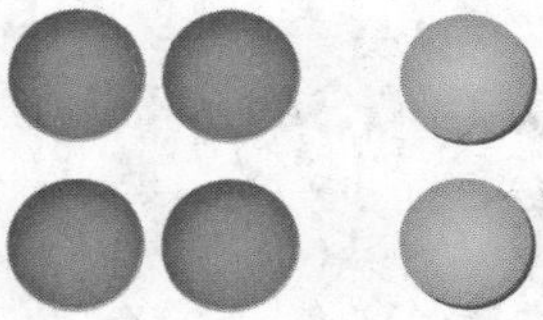

4 + 2 = 6

Teacher Directions: Ideas for Use

- Have students think of words that rhyme with some of the cards.
- Ask pairs of students to sort the cards by the number of syllables on each card.
- Have students draw examples for each card. Have them make drawings that are different from what is shown on the front of each card.

An expression using numbers and the + and = sign.

To join together sets to find the total or sum.

Something that is not a fact. The opposite of true.

The sign used to show having the same value as or is the same as.

The symbol used to show addition.

One of the two parts joined to make the whole.

My Vocabulary Cards

Vocab abc

Processes & Practices

Lesson 1-3

sum

2 + 4 = 6

Lesson 1-13

true

4 + 1 = 5 is true

Lesson 1-2

whole

Part	Part
3	4
Whole	
7	

whole

Lesson 1-4

zero

2 apples   0 apples

Teacher Directions: More Ideas for Use

- Use the blank cards to write your own vocabulary words.
- Have students draw examples for each card. Have them make drawings that are different from what is shown on each card.

Something that is a fact. The opposite of false.

The answer to an addition problem.

A count of no objects.

The sum of two parts.

My Foldable

FOLDABLES® Follow the steps on the back to make your Foldable.

Part	+	Part	=	Whole
1	+	3	=	☐
2	+	6	=	☐
6	+	1	=	☐
3	+	6	=	☐
☐	+	☐	=	10

FOLDABLES® Study Organizer

1

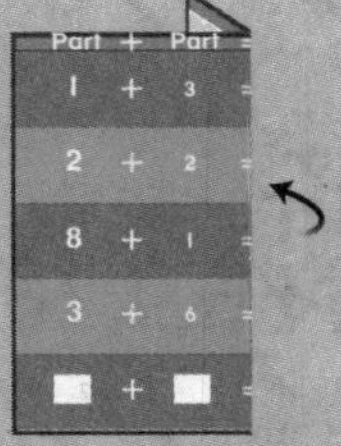

2

3

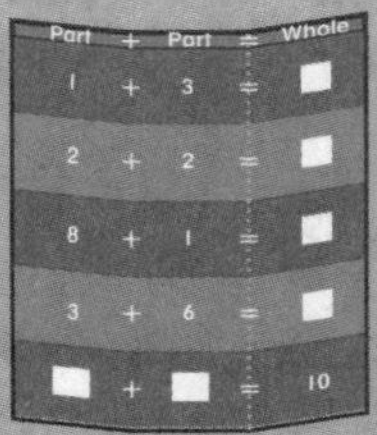

Name ....

Addition Stories

Lesson 1

ESSENTIAL QUESTION
How do you add numbers?

I'm just swinging by to say hi!

Explore and Explain

Watch Tools

Teacher Directions: Use ●● to model. 3 children are swinging. 1 child is on the slide. How many children are at the park in all? Write the number.

See and Show

Processes & Practices

There are 4 ducks in the pond. 4 more ducks walk to the pond. How many ducks are there in all?

8 ducks

Tell a number story. Use counters. Write how many in all.

1.

______ turtles

2.

______ birds

Talk Math Tell how you put groups together.

Name

On My Own

Tell a number story. Use **.**
Write how many in all.

Helpful Hint
Place red counters on the first group and place yellow counters on the second group.

3.

_______ foxes

4.

_______ deer

5.

_______ crabs

Problem Solving

Processes & Practices

Draw a picture to solve.

6. There are 6 gray cats. There are 3 black cats. How many cats are there in all?

_____ cats

7. Ryan has 3 flashlights. He found 2 more. How many flashlights does he have in all?

_____ flashlights

Write Math How do you find how many objects there are in all? Explain.

Name ....

My Homework

Lesson 1
Addition Stories

Homework Helper

eHelp Need help? connectED.mcgraw-hill.com

There are 3 marshmallows on one plate.
There are 2 marshmallows on the other plate.
How many marshmallows are there in all?

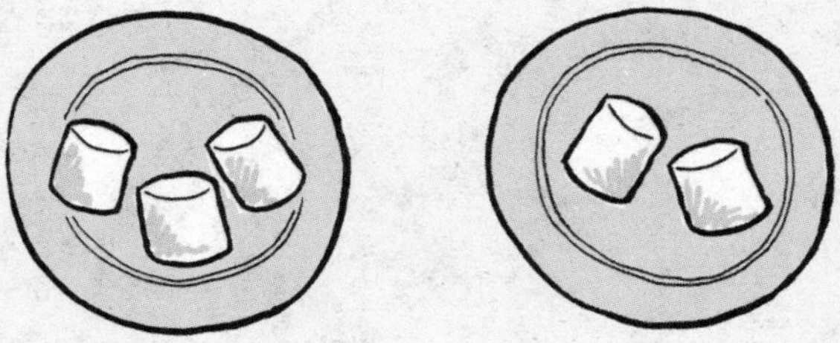

5 marshmallows

Practice

Tell a number story. Write how many there are in all.

1.

_______ sticks

2.

_______ s'mores

Draw a picture to solve.

3. Ella has 5 carrots. She gets 3 more. How many carrots does Ella have in all?

_____ carrots

4. Joe has 2 beans. His mother gives him 2 more. How many beans does Joe have in all?

_____ beans

Test Practice

5. How many peppers are there in all?

5	7	8	9
○	○	○	○

Math at Home Tell addition stories to your child. Have your child use buttons to model the stories.

Name ....

Model Addition

Lesson 2

ESSENTIAL QUESTION
How do you add numbers?

Explore and Explain

Watch Tools

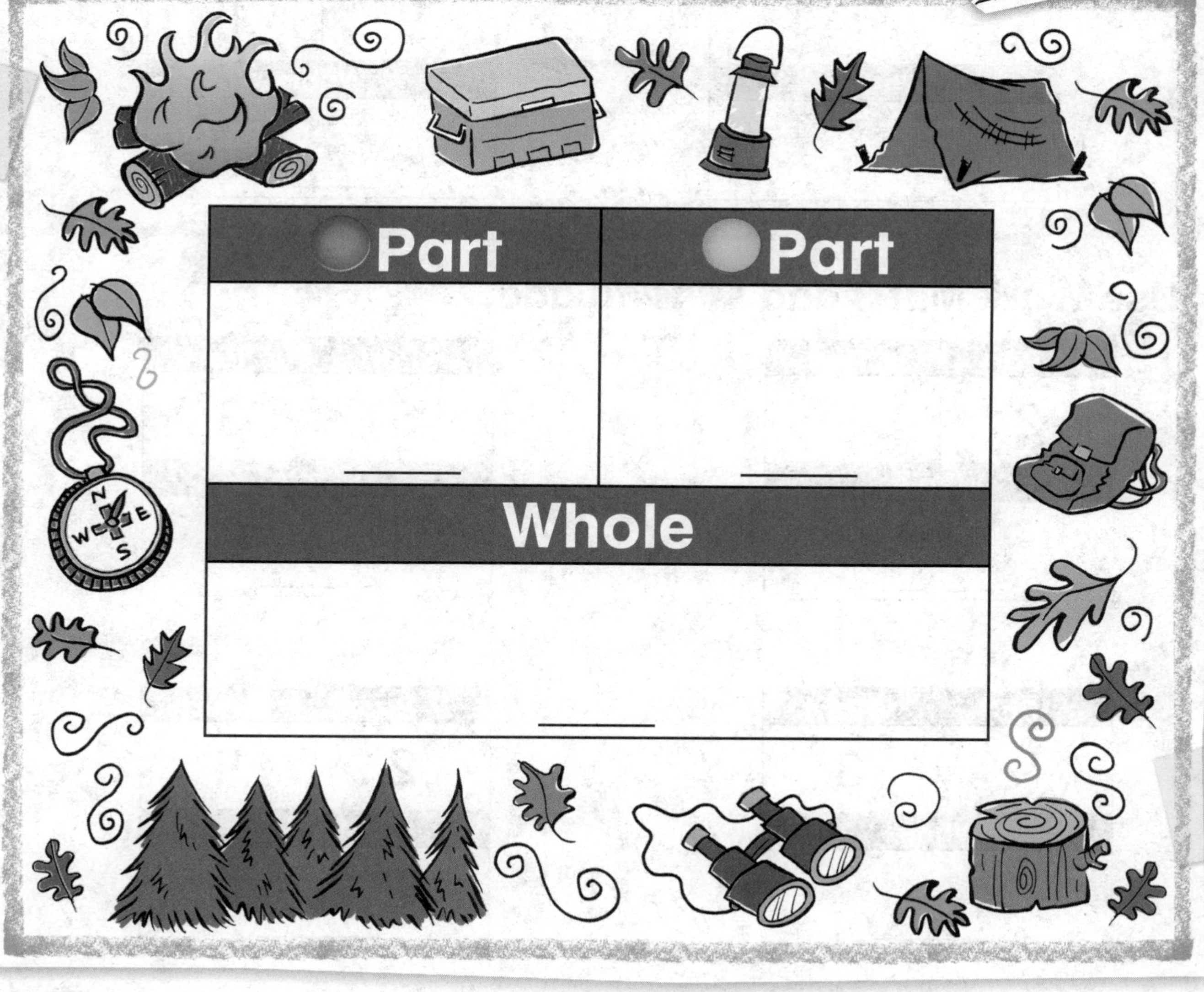

Teacher Directions: Use counters to model. 2 girls bought tents from a store. 1 boy bought a tent from the same store. How many people bought tents in all? Write the numbers. Trace your counters to show the number of people who bought tents.

See and Show

Processes & Practices

To find the **whole**, you **add** the **parts**.

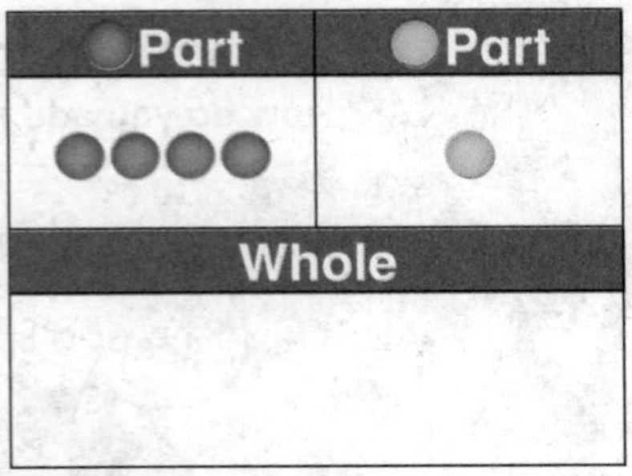

→

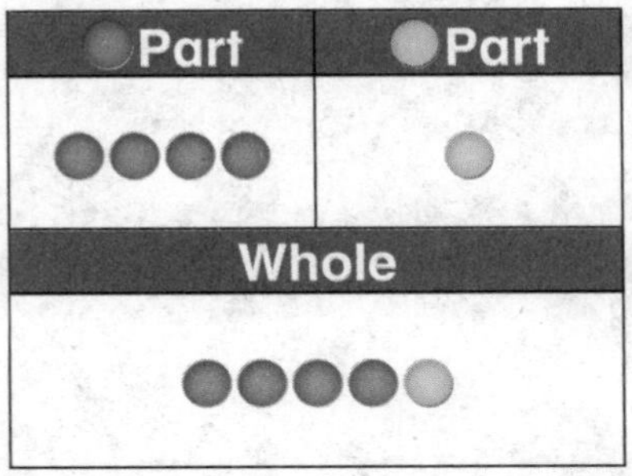

Part	Part
4	1
Whole	

→

Part	Part
4	1
Whole	
5	

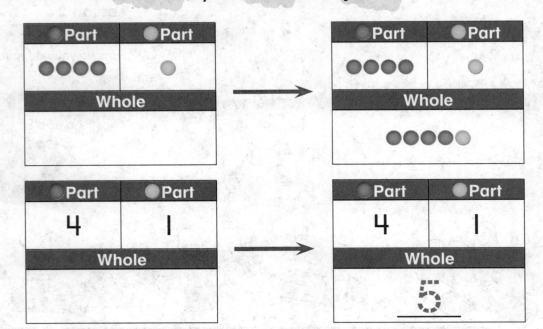

Use Work Mat 3 and ● ● to add.

1.

Part	Part
2	1
Whole	
____	

2.

Part	Part
5	3
Whole	
____	

3.

Part	Part
4	3
Whole	
____	

4.

Part	Part
2	4
Whole	
____	

Talk Math How do you use ● ● to add 7 and 1?

Name ____________________

On My Own

Use Work Mat 3 and (counters) to add.

5.

Part	Part
3	2
Whole	
____	

6.

Part	Part
4	5
Whole	
____	

7.

Part	Part
6	2
Whole	
____	

8.

Part	Part
5	2
Whole	
____	

9.

Part	Part
1	3
Whole	
____	

10.

Part	Part
4	2
Whole	
____	

11.

Part	Part
1	2
Whole	
____	

12.

Part	Part
3	3
Whole	
____	

Problem Solving

Processes & Practices

Use Work Mat 3 and if needed.

13. Cristian saw 6 deer in a field. Camila saw 2 deer in the forest. How many deer did they see in all?

_____ deer

14. Erin picks 6 flowers. Cliff picks 3 flowers. He gives them to Erin. How many flowers does Erin have now?

_____ flowers

Write Math How do you find the whole? Explain.

Name ....

My Homework

Lesson 2
Model Addition

Homework Helper

eHelp

Need help? connectED.mcgraw-hill.com

To find the whole, add the parts.

Part	Part

Whole	

Part	Part
3	4
Whole	
7	

Practice

Use pennies to add. Write the number.

1.

Part	Part
8	1
Whole	
____	

2.

Part	Part
5	2
Whole	
____	

3.

Part	Part
2	3
Whole	
____	

4.

Part	Part
3	5
Whole	
____	

Use pennies to add. Write the number.

5.

Part	Part
1	4
Whole	
____	

6.

Part	Part
5	1
Whole	
____	

7. Stacie caught 4 fish in the morning.
She caught 3 more fish in the afternoon.
How many fish did Stacie catch in all?

_____ fish

Vocabulary Check

Vocab

Complete the sentence.

in all **part**

8. You can add the numbers from each __________ to find the whole.

Math at Home Place 2 red paper circles and 4 yellow paper circles on a table. Have your child count the circles. Ask your child to identify all of the different ways to make 6.

Name ________________________________

Addition Number Sentences

Lesson 3

ESSENTIAL QUESTION
How do you add numbers?

Tag!
You're it!

Explore and Explain

Watch

Tools

____ + ____ = ____
Write your addition sentence here.

Teacher Directions: Use [cube] to model. There are 4 children playing tag. 2 more children join them. How many children are playing tag in all? Trace the cubes. Write the addition number sentence.

See and Show

Processes &Practices

You can write an addition number sentence.

See

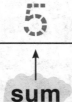

Say 3 **plus** 2 **equals** 5

sum

Write 3 + 2 = 5

3 + 2 = 5 is an **addition number sentence**.

Write an addition number sentence.

1. 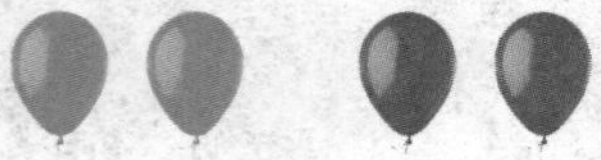

___ ◯ ___ ◯ ___

2. 

___ ◯ ___ ◯ ___

3. 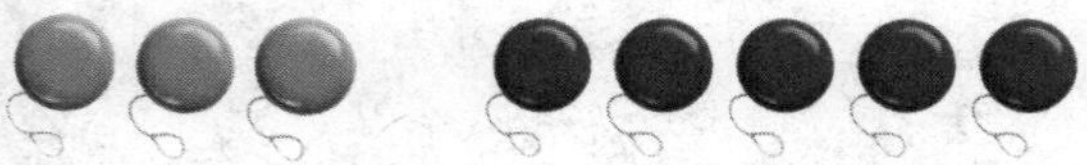

___ ◯ ___ ◯ ___

4. 

___ ◯ ___ ◯ ___

Talk Math What does the symbol + mean?

Name ....

On My Own

Write an addition number sentence.

5. 

_____ ○ _____ ○ _____

6. 

_____ ○ _____ ○ _____

7. 

_____ ○ _____ ○ _____

8. 

_____ ○ _____ ○ _____

9. 

_____ ○ _____ ○ _____

10. 

_____ ○ _____ ○ _____

11. 

_____ ○ _____ ○ _____

12. 

_____ ○ _____ ○ _____

Problem Solving

Processes & Practices

13. There are 2 dogs playing. 3 more dogs join them. How many dogs are playing in all?

____ ◯ ____ ◯ ____ dogs

14. Suzy sees 2 bees flying around a flower. She sees 4 more bees on the flower. How many bees are there in all?

____ ◯ ____ ◯ ____ bees

Write Math What does = mean?

Name ....

My Homework

Lesson 3
Addition Number Sentences

Homework Helper

eHelp **Need help?** **connectED.mcgraw-hill.com**

You can write an addition number sentence.

2 + 4 = 6

Practice

Write an addition number sentence.

1.

____ ◯ ____ ◯ ____

2.

____ ◯ ____ ◯ ____

3.

____ ◯ ____ ◯ ____

4.

____ ◯ ____ ◯ ____

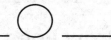

Write an addition number sentence.

5. 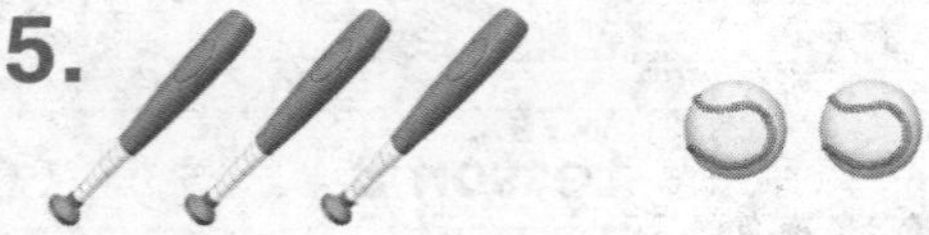

____ ◯ ____ ◯ ____

6. 

____ ◯ ____ ◯ ____

7. There are 5 cats at the park. 2 more join them. How many cats are there in all?

____ ◯ ____ ◯ ____ cats

8. There are 4 squirrels on a tree. 2 more squirrels join them. How many squirrels are there in all?

____ ◯ ____ ◯ ____ squirrels

Vocabulary Check

Draw lines to match.

9. **addition number sentence**	The answer to an addition problem.
10. **sum**	$4 + 5 = 9$

Math at Home Create addition stories using cans of fruit or vegetables. Have your child write addition number sentences for the stories.

Name ....

Add 0

Lesson 4

ESSENTIAL QUESTION
How do you add numbers?

You're sweet!

Explore and Explain

Watch Tools

Lemonade for Sale
____ + ____ = ____
Write your addition sentence here.

Teacher Directions: Use [cube icon] to model. 7 people bought a glass of lemonade in the afternoon. 0 people bought a glass of lemonade in the evening. How many glasses of lemonade were sold in all? Write the addition number sentence.

See and Show

Processes & Practices

When you add **zero** to a number, the sum is the same as the number.

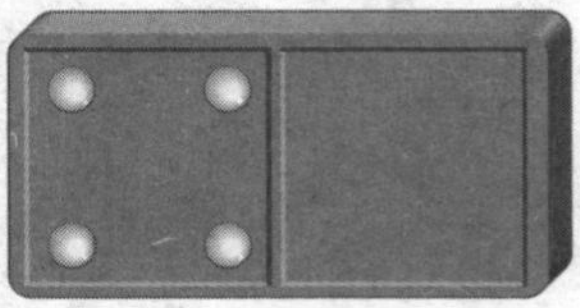

4 + **0** = 4
sum

When you add a number to zero, the sum is the same as the number.

0 + 2 = 2
sum

Find each sum.

1\.

0 + 8 = ______

2\.

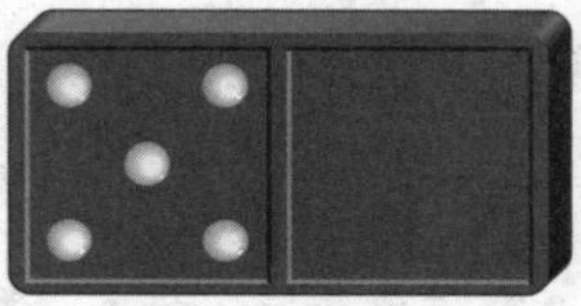

5 + 0 = ______

3\.

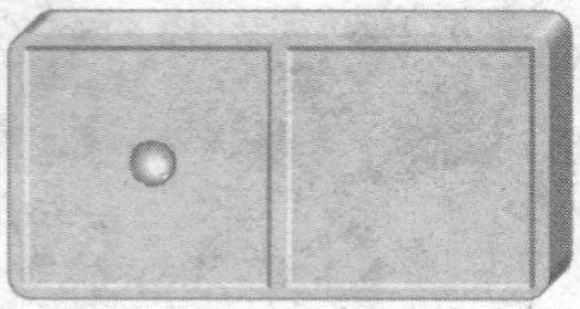

1 + 0 = ______

4\.

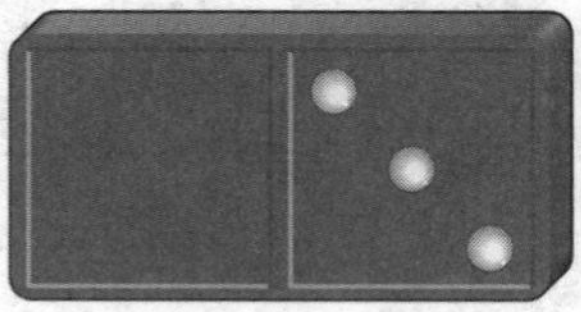

0 + 3 = ______

Talk Math What happens when you add zero to a number? Explain.

Name ________________________________

On My Own

Find each sum.

5.

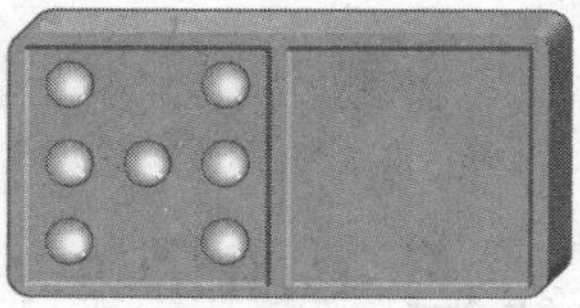

7 + 0 = ______

6.

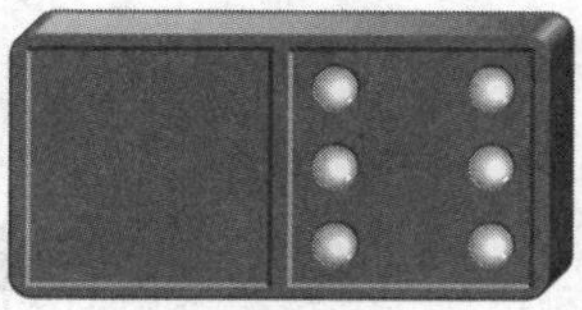

0 + 6 = ______

7. 3 + 1 = ______

8. 8 + 0 = ______

9. 3 + 0 = ______

10. 2 + 3 = ______

11. 0 + 9 = ______

12. 0 + 5 = ______

13. 1 + 3 = ______

14. 0 + 2 = ______

15. 4 + 2 = ______

Problem Solving

Processes & Practices

16. Jackson has 4 canoe paddles. Ian has 0 paddles. How many paddles do they have altogether?

______ paddles

17. Grayson ate 2 hot dogs. His friends ate 4 hot dogs. How many hot dogs did they eat altogether?

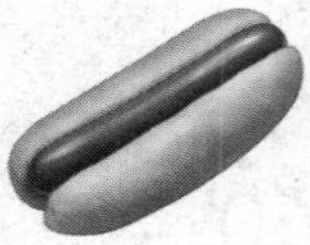

______ hot dogs

HOT Problem Adrian adds 6 + 0 like this. Tell why Adrian is wrong. Make it right.

$6 + 0 = 0$

____

____

____

Name ....

My Homework

Lesson 4
Add 0

Homework Helper

eHelp **Need help?** connectED.mcgraw-hill.com

When you add zero, you add none.

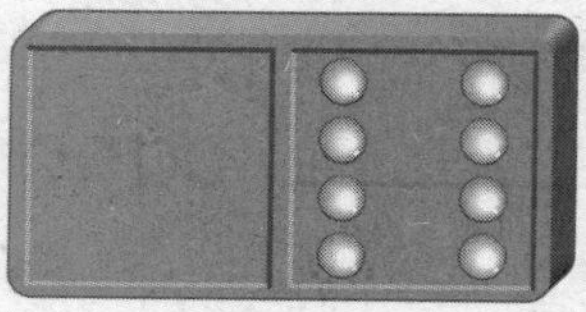

0 + 8 = 8

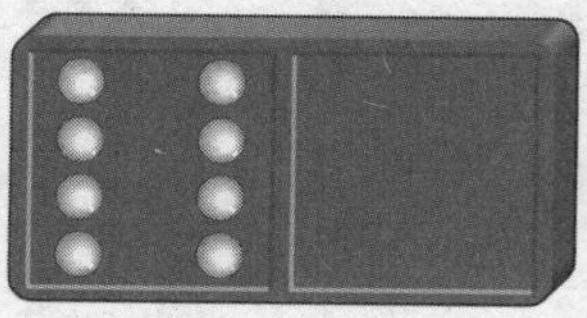

8 + 0 = 8

Practice

Find each sum.

1.

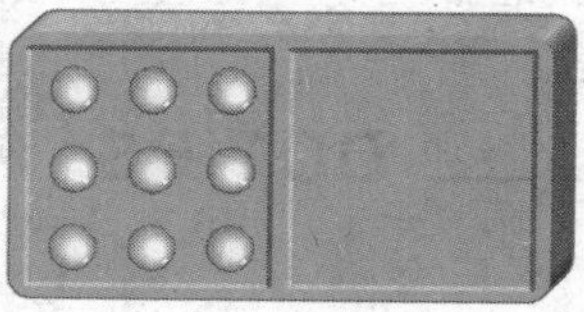

9 + 0 = ______

2.

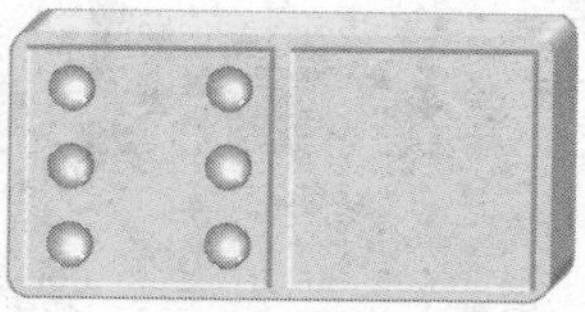

6 + 0 = ______

3.

0 + 4 = ______

4.

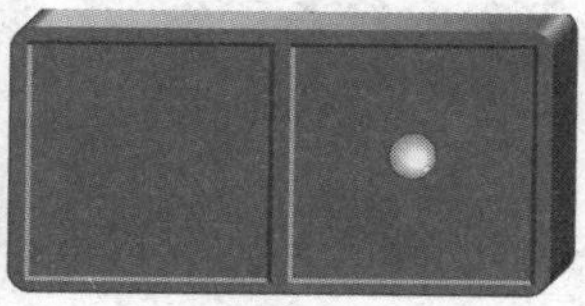

0 + 1 = ______

Find each sum.

5. 0 + 5 = ______

6. 3 + 5 = ______

7. 2 + 2 = ______

8. 0 + 9 = ______

9. There are 8 canoes on the lake.
There are 0 canoes on the land.
How many canoes are there in all?

______ canoes

10. There are 5 apples in a bag. There are 0 apples in another bag. How many apples are there altogether?

______ apples

Vocabulary Check

Vocab

Circle the correct number.

11. **zero** 1 6 0

Math at Home Hold some cereal in one hand. Hold no cereal in another hand. Hold both hands out to your child. Ask your child to tell you which hand has zero pieces of cereal in it. Have your child add the cereal in each hand. Have your child say how many pieces of cereal there are in all.

Name

Check My Progress

Vocabulary Check

Draw lines to match.

1. =	**addition number sentence**
2. +	**equals**
3. 2 + 3 = 5	**plus**
4. 0	**zero**

Circle the correct answer.

5. When you ______ numbers together, you find the sum.

zero **add**

6. You can find the whole by adding the ______.

parts **sum**

7. Add two parts to find the ______.

whole **equals**

Concept Check

Add. Write the number.

8.

Part	Part
2	4
Whole	
____	

9.

Part	Part
3	5
Whole	
____	

Write an addition number sentence.

10. 

____ ◯ ____ ◯ ____

11. There are 4 red apples and 2 green apples in a bag. How many apples are there in all?

____ ◯ ____ ◯ ____ apples

Test Practice

12. Find the addition number sentence that matches.

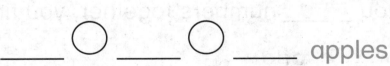

$5 + 2 = 7$ ◯

$5 + 1 = 6$ ◯

$4 + 2 = 6$ ◯

$4 + 3 = 7$ ◯

Name ____________________

Vertical Addition

Lesson 5

ESSENTIAL QUESTION
How do you add numbers?

Explore and Explain

Watch Tools

____ + ____ = ____

$$\begin{array}{r} \square \\ + \ \square \\ \hline \square \end{array}$$

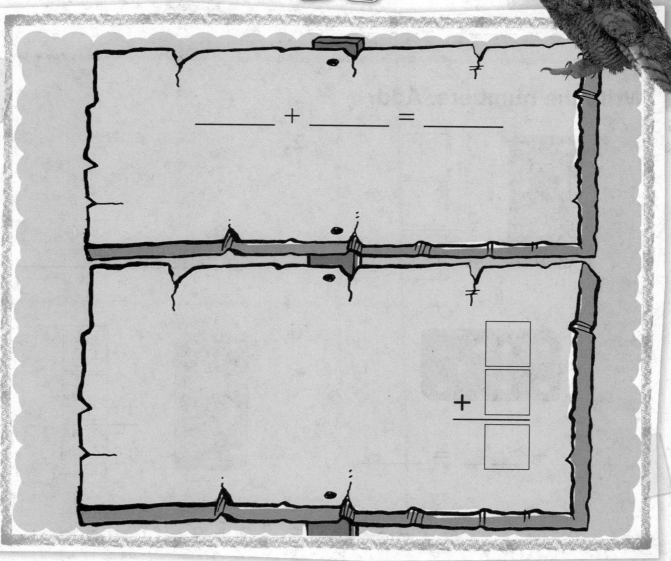

Teacher Directions: Use ●● to model. Show 2 + 1 on each sign. Trace the counters. Write the addition number sentences.

See and Show

Processes & Practices

You can add across. You can add down. The sum is the same if the numbers added are the same.

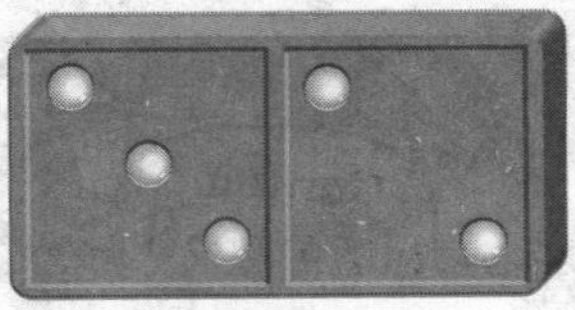

3 + 2 = 5

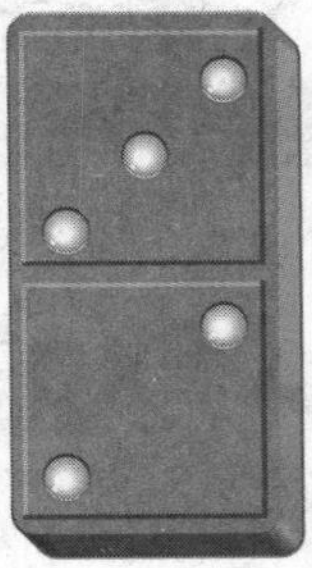

$$\begin{array}{r} 3 \\ +\ 2 \\ \hline 5 \end{array}$$

Write the numbers. Add.

1\. 

$$\begin{array}{r} \square \\ +\ \square \\ \hline \square \end{array}$$

2\. 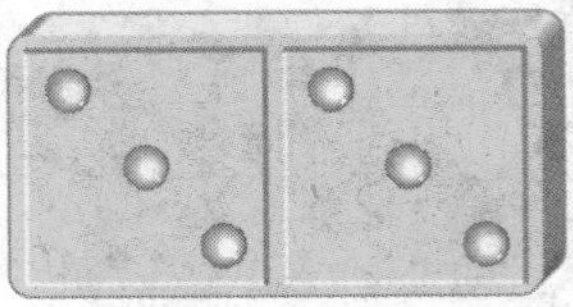

____ + ____ = ____

3\. 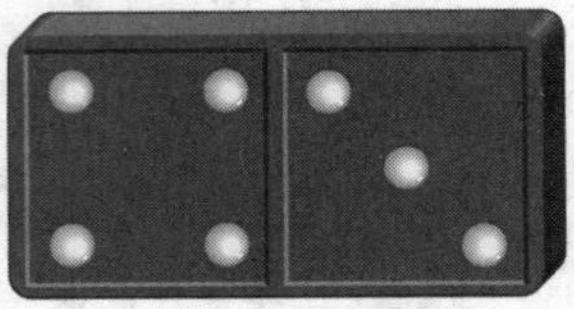

____ + ____ = ____

4\. 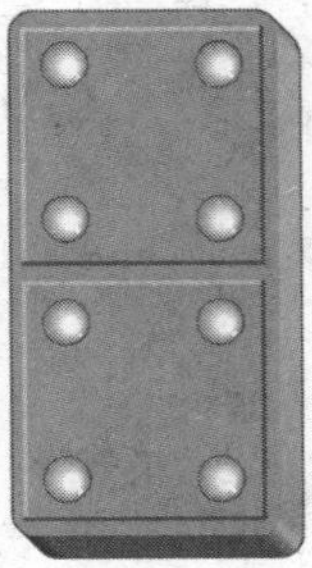

$$\begin{array}{r} \square \\ +\ \square \\ \hline \square \end{array}$$

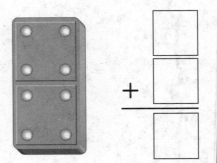

Talk Math You know that 5 + 3 = 8. If you add down, what is the sum? Explain.

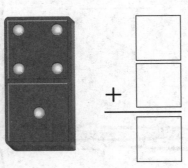

Name ....

On My Own

Write the numbers. Add.

5.

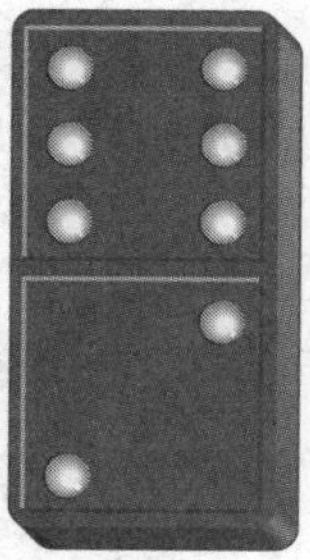

☐
+ ☐
☐

6.

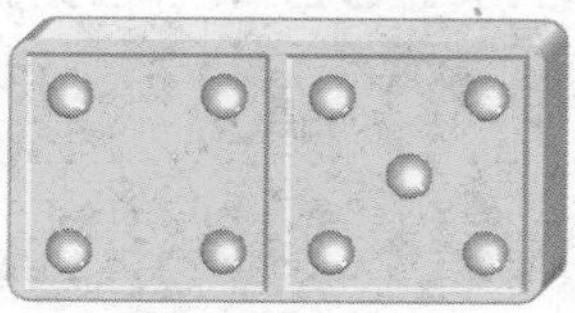

_____ + _____ = _____

7.

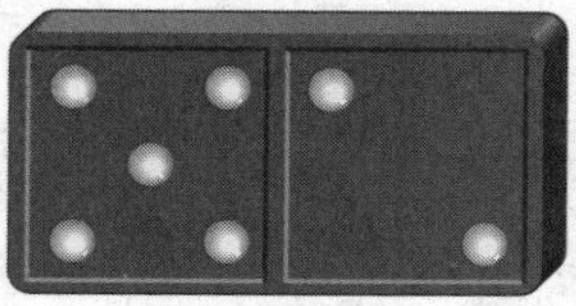

_____ + _____ = _____

8.

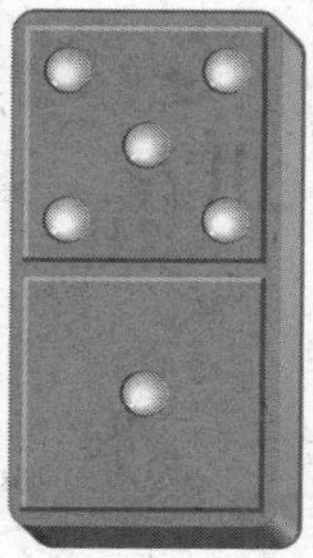

☐
+ ☐
☐

9.

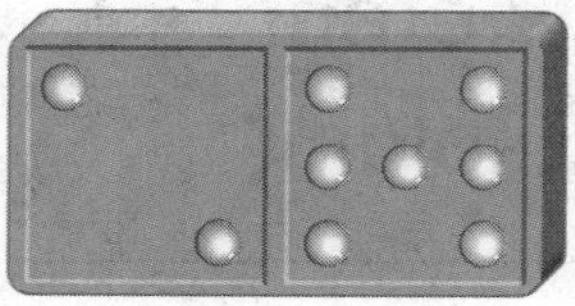

_____ + _____ = _____

10.

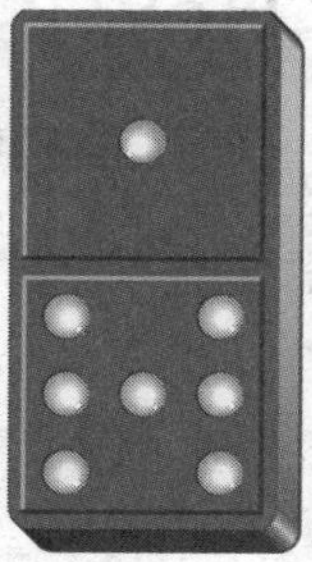

☐
+ ☐
☐

Add.

11. 2 + 6 = _____

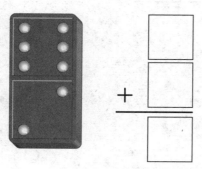

12. $\begin{array}{r} 4 \\ +\ 5 \\ \hline \end{array}$

13. 1 + 3 = _____

Real World

Problem Solving

Processes & Practices

14. Brian saw 5 foxes in a field. He saw 3 other foxes in the woods. How many foxes did Brian see in all?

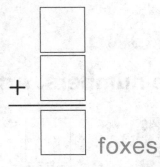

$$\begin{array}{r} \square \\ +\ \square \\ \hline \square \end{array}$$ foxes

15. Nate found 3 bugs. Pablo found 2 bugs. How many bugs did they find in all?

$$\begin{array}{r} \square \\ +\ \square \\ \hline \square \end{array}$$ bugs

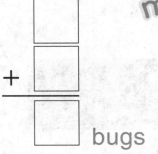

Write Math How is adding down different from adding across? Explain.

___

___

___

Name ....

My Homework

Lesson 5
Vertical Addition

Homework Helper

eHelp Need help? connectED.mcgraw-hill.com

You can add across or you can add down.

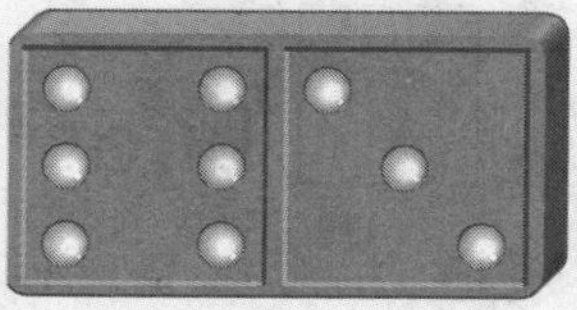

$6 + 3 = 9$

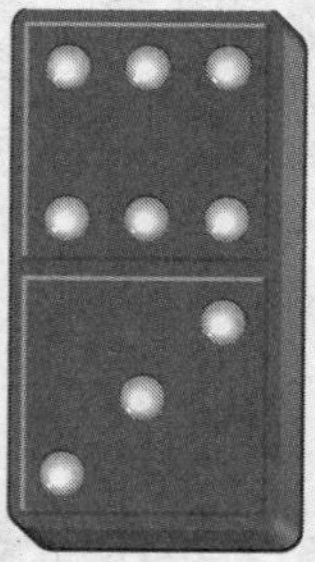

$$\begin{array}{r} 6 \\ +\ 3 \\ \hline 9 \end{array}$$

Practice

Write the numbers. Add.

1.

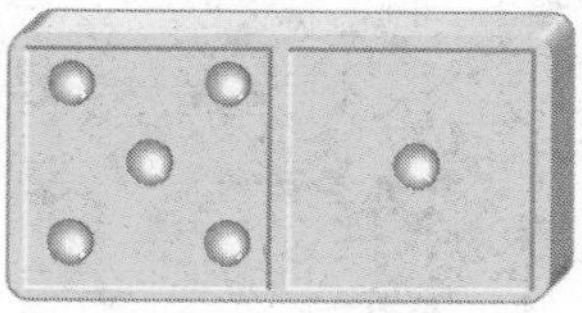

____ + ____ = ____

2.

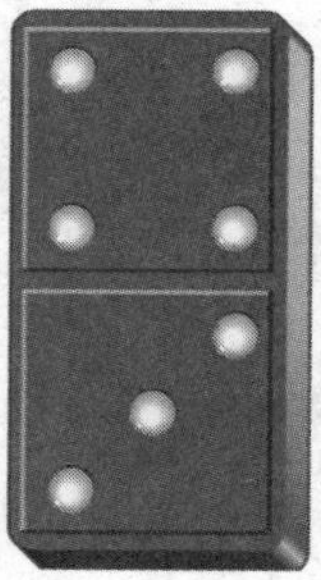

$$\begin{array}{r} \square \\ +\ \square \\ \hline \square \end{array}$$

Add.

3. $2 + 6 =$ ____

4. $2 + 2 =$ ____

5. $1 + 4 =$ ____

Add.

6. $\begin{array}{r} 2 \\ +\ 7 \\ \hline \end{array}$

7. 2 + 3 = ____

8. $\begin{array}{r} 1 \\ +\ 6 \\ \hline \end{array}$

9. There are 2 birds in a nest. 2 more birds fly to the nest. How many birds are there in all?

$\begin{array}{r} \square \\ +\ \square \\ \hline \square \end{array}$ birds

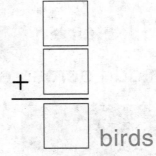

10. There are 5 children hiking. Then 3 more children join them. How many children are hiking in all?

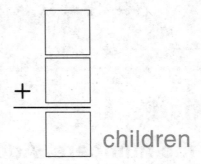

$\begin{array}{r} \square \\ +\ \square \\ \hline \square \end{array}$ children

Test Practice

11. 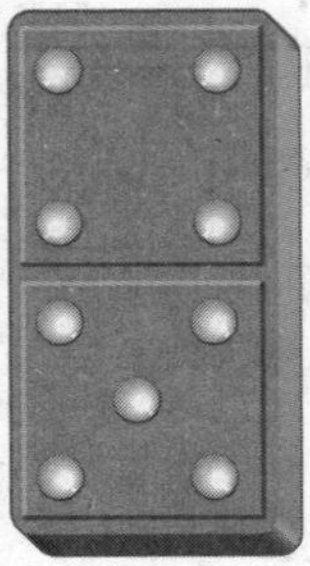

$\begin{array}{r} 4 \\ +\ 5 \\ \hline \square \end{array}$

1	3	8	9
○	○	○	○

Math at Home Give your child an addition number sentence. Have your child show how to add across and down.

Name

Problem Solving

STRATEGY: Write a Number Sentence

Lesson 6

ESSENTIAL QUESTION
How do you add numbers?

2 children are fishing.
4 more children join them.
How many children are fishing in all?

1 Understand Underline what you know. Circle what you need to find.

2 Plan How will I solve the problem?

3 Solve I will write a number sentence.

2 + 4 = 6

6 children are fishing in all.

4 Check Is my answer reasonable? Explain.

Practice the Strategy

Cassie saw 2 elk in a field.
Marta saw 5 elk in the forest.
How many elk do they see altogether?

1 Understand Underline what you know.
Circle what you need to find.

2 Plan How will I solve the problem?

3 Solve I will...

____ ◯ ____ ◯ ____

They see ____ elk altogether.

4 Check Is my answer reasonable? Explain.

Name ....

Apply the Strategy

Write an addition number sentence to solve.

1. Leon has 5 cards. Trey has 4 cards. How many cards do they have in all?

____ ◯ ____ ◯ ____ cards

2. Nicki has 6 stickers. She was given 2 more stickers. How many stickers does she have now?

____ ◯ ____ ◯ ____ stickers

3. Isi saw 6 cars going down the road. Jamaal saw 3 cars. How many cars did Isi and Jamaal see?

____ ◯ ____ ◯ ____ cars

Review the Strategies

Choose a strategy
- Write a number sentence.
- Make a table.
- Act it out.

4. Jayla and Will each have 4 fish. How many total fish do Jayla and Will have?

_____ fish

5. Deon has 4 jump ropes. Karen has 3 jump ropes. How many jump ropes do they have in all?

_____ jump ropes

6. There are 5 yellow beads and 4 red beads on a necklace. How many beads are on the necklace in all?

_____ beads

Name ....

My Homework

Lesson 6
Problem Solving: Write a Number Sentence

Homework Helper eHelp **Need help?** connectED.mcgraw-hill.com

There are 2 birds singing.
4 more birds begin singing.
How many birds are singing in all?

1 Understand Underline what you know.
Circle what you need to find.

2 Plan How will I solve the problem?

3 Solve I will write a number sentence.

2 + 4 = 6 birds

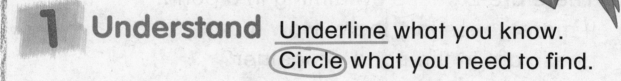

4 Check Is my answer reasonable?

Problem Solving

Underline what you know. Circle what you need to find. Write an addition number sentence.

1. Andrew saw 3 rabbits. Tia saw 6 other rabbits. How many rabbits did they see in all?

   ____ ◯ ____ ◯ ____ rabbits

2. There are 2 geese swimming in a pond. 4 more geese join them. How many geese are in the pond altogether?

   ____ ◯ ____ ◯ ____ geese

3. Ben saw 5 raccoons. Conner saw 2 other raccoons. How many raccoons did they see in all?

   ____ ◯ ____ ◯ ____ raccoons

Math at Home Take advantage of problem-solving opportunities during daily routines such as riding in the car, bedtime, doing laundry, putting away groceries, planning schedules, and so on.

Name ....

Ways to Make 4 and 5

Lesson 7

ESSENTIAL QUESTION
How do you add numbers?

Mmm!!!

Explore and Explain

Watch Tools

_____ + _____ = _____
Write your addition sentence here.

Teacher Directions: Use ●● to model. Mia's dad put 1 piece of wood on the fire. Then he put 3 more pieces on it. How many pieces of wood are on the fire in all? Trace the counters you used. Write the addition number sentence.

See and Show

Processes & Practices

There are many ways to make a sum of 4 and 5.

2 + 2 = 4

3 + 1 = 4

2 + 3 = 5

1 + 4 = 5

Use Work Mat 3 and ●● to show different ways to make a sum of 4. Color the ○. Write the numbers.

Ways to Make 4

1. ●●●● ____ + ____ = 4

2. ○○○○ ____ + ____ = 4

3. ○○○○ ____ + ____ = 4

Talk Math What is another way to make 4?

Name ____________________

On My Own

Use Work Mat 3 and ●● to show different ways to make a sum of 5. Color the ◯. Write the numbers.

Ways to Make 5

4. ◯ ◯ ◯ ◯ ◯ ____ + ____ = 5

5. ◯ ◯ ◯ ◯ ◯ ____ + ____ = 5

6. ◯ ◯ ◯ ◯ ◯ ____ + ____ = 5

7. ◯ ◯ ◯ ◯ ◯ ____ + ____ = 5

Add.

8. 1 + 4 = ____

9. 3 + 1 = ____

10. 3 + 2 = ____

11. 2 + 2 = ____

12. $\begin{array}{r} 1 \\ +\ 4 \\ \hline \end{array}$

13. $\begin{array}{r} 4 \\ +\ 0 \\ \hline \end{array}$

14. $\begin{array}{r} 0 \\ +\ 5 \\ \hline \end{array}$

Problem Solving

Real World!

Processes & Practices

Write an addition number sentence.

15. Kylie has 3 maps. Taye has 2 maps. How many maps do they have in all?

_____ + _____ = _____ maps

16. Kate saw 1 turkey. Malik saw 3 other turkeys. How many turkeys did they see in all?

_____ + _____ = _____ turkeys

Write Math Is there more than one way to make 5? Explain.

Name ....

My Homework

Lesson 7

Ways to Make 4 and 5

Homework Helper

eHelp **Need help?** **connectED.mcgraw-hill.com**

There are different ways to make a sum of 4 or 5.

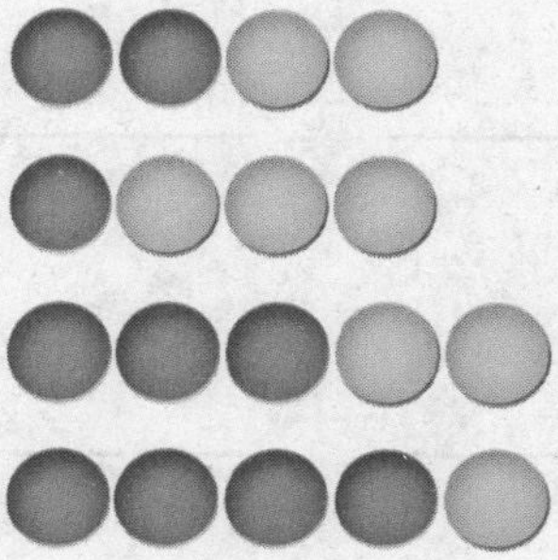

2 + 2 = 4

1 + 3 = 4

3 + 2 = 5

4 + 1 = 5

Practice

Write different ways to make 4.

1. 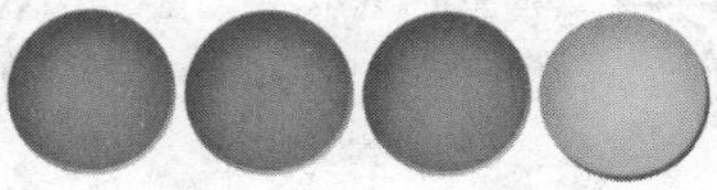 ______ + ______ = 4

2.  ______ + ______ = 4

3. 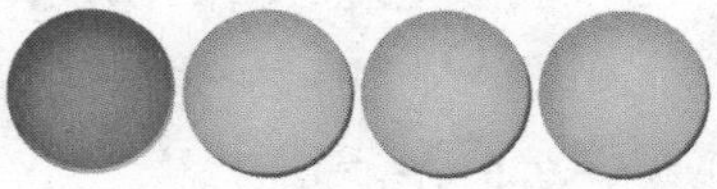 ______ + ______ = 4

4. 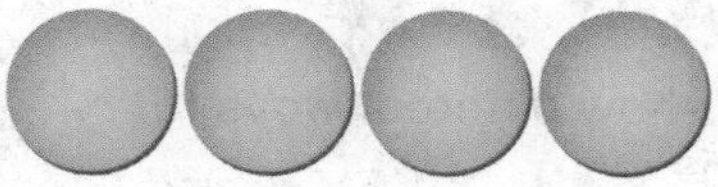 ______ + ______ = 4

Write different ways to make 5.

5.  _____ + _____ = 5

6.  _____ + _____ = 5

7.  _____ + _____ = 5

8.  _____ + _____ = 5

9.  _____ + _____ = 5

10. Jose saw 3 green frogs and 1 red frog. How many frogs did he see in all?

_____ frogs

Test Practice

11. How many rainbows are there in all?

3	4	5	6
○	○	○	○

Math at Home Give your child five objects. Have your child show different ways to make 5.

Name

Ways to Make 6 and 7

Lesson 8

ESSENTIAL QUESTION
How do you add numbers?

I love camping!

Explore and Explain

Watch

Tools

+ =
Write your addition sentence here.

Teacher Directions: Use counters to model. Find as many ways as you can to make 6 and 7. Trace counters above the tent to show one of those ways. Write the addition number sentence.

See and Show

Processes & Practices

There are many ways to make a sum of 6 and 7.

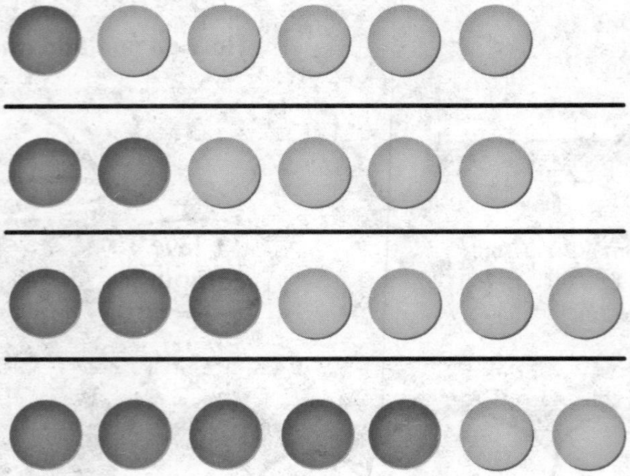

1 + 5 = 6

2 + 4 = 6

3 + 4 = 7

5 + 2 = 7

Use Work Mat 3 and to show different ways to make a sum of 6. Write the numbers.

Ways to Make 6

1. ______ + ______ = 6

2. ______ + ______ = 6

3. ______ + ______ = 6

4. ______ + ______ = 6

Helpful Hint
Think of all of the different ways to make 6.

Talk Math Is 5 + 1 the same as 4 + 2? Explain.

Name

On My Own

Use Work Mat 3 and ●● to show different ways to make a sum of 7. Write the numbers.

Ways to Make 7

5. ____ + ____ = 7

6. ____ + ____ = 7

7. ____ + ____ = 7

8. ____ + ____ = 7

9. ____ + ____ = 7

10. ____ + ____ = 7

Add.

11. 4 + 2 = ____

12. 3 + 4 = ____

13. 7 + 0 = ____

14. 3 + 3 = ____

15. 4 + 3 = ____

16. 5 + 1 = ____

17. $\begin{array}{r} 1 \\ +\ 6 \\ \hline \end{array}$

18. $\begin{array}{r} 0 \\ +\ 6 \\ \hline \end{array}$

19. $\begin{array}{r} 5 \\ +\ 2 \\ \hline \end{array}$

Problem Solving

Processes & Practices

Write an addition number sentence.

20. Victoria caught 5 fish. Her brother caught 2 fish. How many fish did they catch altogether?

_____ + _____ = _____ fish

21. There are 4 dogs swimming in a pond. 3 more dogs join them. How many dogs are swimming in the pond in all?

_____ + _____ = _____ dogs

HOT Problem Mason added 3 + 3 like this. Tell why Mason is wrong. Make it right.

3 + 3 = 7

Name ______________________

My Homework

Lesson 8
Ways to Make 6 and 7

Homework Helper

eHelp Need help? connectED.mcgraw-hill.com

There are many ways to make sums of 6 and 7.

4 + 2 = 6

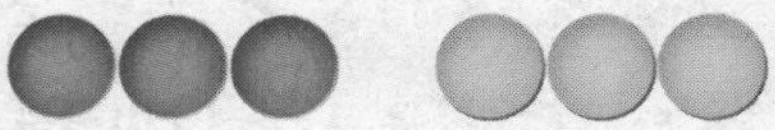

3 + 3 = 6

5 + 2 = 7

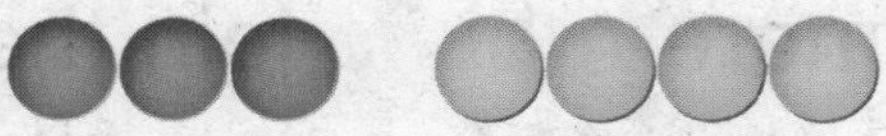

3 + 4 = 7

Practice

Write different ways to make 6 and 7.

Ways to Make 6 or 7

1. ______ + ______ = 6
2. ______ + ______ = 6
3. ______ + ______ = 6
4. ______ + ______ = 6
5. ______ + ______ = 7
6. ______ + ______ = 7
7. ______ + ______ = 7
8. ______ + ______ = 7

Add.

9. $\begin{array}{r} 6 \\ +\ 0 \\ \hline \end{array}$	**10.** $\begin{array}{r} 5 \\ +\ 2 \\ \hline \end{array}$	**11.** $\begin{array}{r} 4 \\ +\ 3 \\ \hline \end{array}$
12. $\begin{array}{r} 1 \\ +\ 6 \\ \hline \end{array}$	**13.** $\begin{array}{r} 3 \\ +\ 3 \\ \hline \end{array}$	**14.** $\begin{array}{r} 2 \\ +\ 4 \\ \hline \end{array}$

15. There are 2 rabbits eating. 5 more rabbits join them. How many rabbits are eating in all?

_____ + _____ = _____ rabbits

Test Practice

16. How many trees are there in all?

4	5	6	7
○	○	○	○

Math at Home Give your child 7 objects. Then have your child show different ways to make two groups that show 7 in all.

Name ....

Ways to Make 8

Lesson 9

ESSENTIAL QUESTION
How do you add numbers?

Jump in!

Explore and Explain

Watch Tools

____ + ____ = ____
Write your addition sentence here.

Teacher Directions: Use [counters] to model. There are 5 people swimming. 3 more people join them. How many people are swimming in all? Write the addition number sentence.

See and Show

There are many ways to make sums of 8.

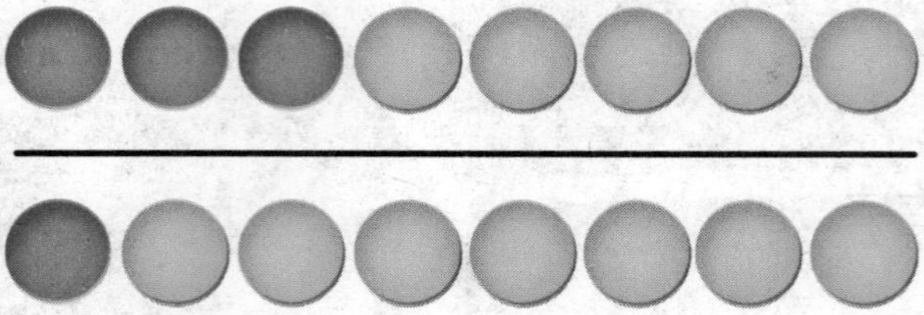

3 + 5 = 8

1 + 7 = 8

Use Work Mat 3 and counters to show different ways to make a sum of 8. Write the numbers.

Ways to Make 8

1. ______ + ______ = 8
2. ______ + ______ = 8
3. ______ + ______ = 8
4. ______ + ______ = 8
5. ______ + ______ = 8
6. ______ + ______ = 8

Add.

7. 4 + 4 = ______
8. 6 + 2 = ______
9. 8 + 0 = ______
10. 7 + 1 = ______

Talk Math How could you use cubes to show ways to make 8?

Name ....

On My Own

Use Work Mat 3 and . Add.

11. 6 + 2 = ______

12. 0 + 3 = ______

13. 2 + 5 = ______

14. 2 + 6 = ______

15. 4 + 4 = ______

16. 0 + 7 = ______

17. 1 + 7 = ______

18. 3 + 5 = ______

19. 2 + 6 = ______

20. 8 + 0 = ______

21. 4 + 2 = ______

22. 7 + 1 = ______

23. $\begin{array}{r} 4 \\ +\ 3 \\ \hline \end{array}$

24. $\begin{array}{r} 4 \\ +\ 4 \\ \hline \end{array}$

25. $\begin{array}{r} 3 \\ +\ 1 \\ \hline \end{array}$

26. $\begin{array}{r} 2 \\ +\ 3 \\ \hline \end{array}$

27. $\begin{array}{r} 6 \\ +\ 1 \\ \hline \end{array}$

28. $\begin{array}{r} 5 \\ +\ 3 \\ \hline \end{array}$

Problem Solving

Real World

Processes & Practices

Write an addition number sentence.

29. Brice saw 5 foxes. His friend saw 3 other foxes. How many foxes did they see in all?

_____ + _____ = _____ foxes

30. There are 4 chipmunks on a tree. 4 more chipmunks join them. How many chipmunks are on the tree now?

_____ + _____ = _____ chipmunks

HOT Problem A baker sold 4 muffins in the morning. She sold 2 muffins later that day. The answer is 6 muffins. What is the question?

____

____

____

Name ____________________

My Homework

Lesson 9
Ways to Make 8

Homework Helper

eHelp **Need help?** connectED.mcgraw-hill.com

There are many ways to make a sum of 8.

5 + 3 = 8

2 + 6 = 8

Practice

Write different ways to make 8.

1. ____ + ____ = 8

2. ____ + ____ = 8

3. ____ + ____ = 8

4. ____ + ____ = 8

5. ____ + ____ = 8

6. ____ + ____ = 8

Add.

7. 3 + 5 = ____

8. 8 + 0 = ____

Add.

9. $\begin{array}{r} 6 \\ +\ 2 \\ \hline \end{array}$

10. $\begin{array}{r} 3 \\ +\ 5 \\ \hline \end{array}$

11. $\begin{array}{r} 4 \\ +\ 3 \\ \hline \end{array}$

12. $\begin{array}{r} 4 \\ +\ 4 \\ \hline \end{array}$

13. $\begin{array}{r} 6 \\ +\ 1 \\ \hline \end{array}$

14. $\begin{array}{r} 1 \\ +\ 7 \\ \hline \end{array}$

15. 5 people went canoeing. 2 more people joined them. How many people went canoeing in all?

______ + ______ = ______ people

Test Practice

16. Which has a sum of 8?

6 + 1	3 + 4	0 + 5	3 + 5
○	○	○	○

Math at Home Give your child 8 objects. Then have your child show different ways to make a sum of 8.

Name ____________________

Check My Progress

Vocabulary Check

Circle the correct answer.

add **addition number sentence** **sum**

1. A ________ is an answer to an addition problem.

Concept Check

Add. Write the numbers.

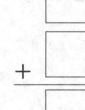

2. 

____ + ____ = ____

3. 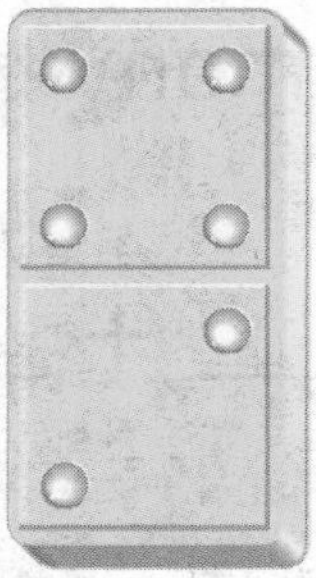

$$\begin{array}{r} \square \\ +\ \square \\ \hline \square \end{array}$$

4. 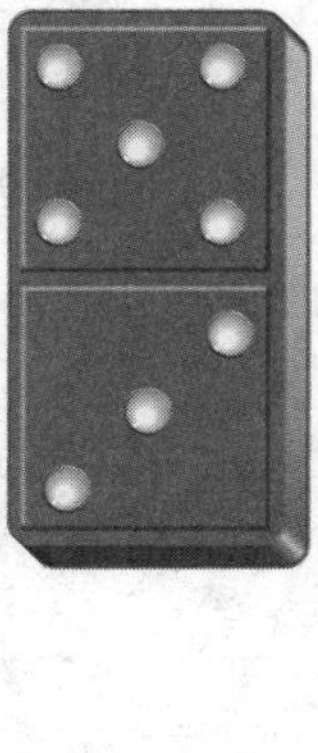

$$\begin{array}{r} \square \\ +\ \square \\ \hline \square \end{array}$$

5. 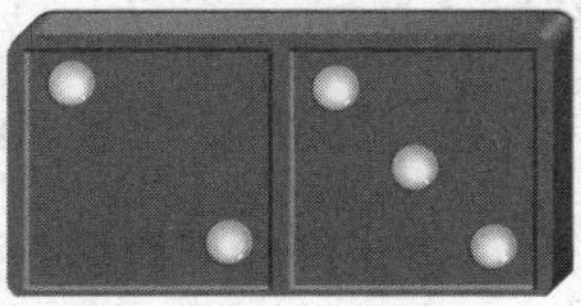

____ + ____ = ____

Add.

6. $2 + 3 =$ ______

7. $2 + 2 =$ ______

8. $\begin{array}{r} 5 \\ +\ 3 \\ \hline \end{array}$

9. $\begin{array}{r} 6 \\ +\ 2 \\ \hline \end{array}$

10. $\begin{array}{r} 1 \\ +\ 5 \\ \hline \end{array}$

11. $\begin{array}{r} 3 \\ +\ 4 \\ \hline \end{array}$

12. $\begin{array}{r} 1 \\ +\ 7 \\ \hline \end{array}$

13. $\begin{array}{r} 1 \\ +\ 3 \\ \hline \end{array}$

Write an addition number sentence to solve.

14. There are 6 birds flying together. There is 1 other bird sitting on a tree branch. How many birds are there in all?

______ + ______ = ______ birds

Test Practice

15. There are 2 beavers swimming in a river. 2 more beavers join them. How many beavers are there in all?

○ 3 beavers   ○ 4 beavers   ○ 5 beavers   ○ 6 beavers

Name ______________________________

Ways to Make 9

Lesson 10

ESSENTIAL QUESTION
How do you add numbers?

It's a pretty night!

Explore and Explain

Watch Tools

___ + ___ = ___
Write your addition sentence here.

Teacher Directions: Use two crayons to color the stars to show a way to make 9. Write the addition number sentence.

See and Show

Processes & Practices

There are many ways to make a sum of 9.

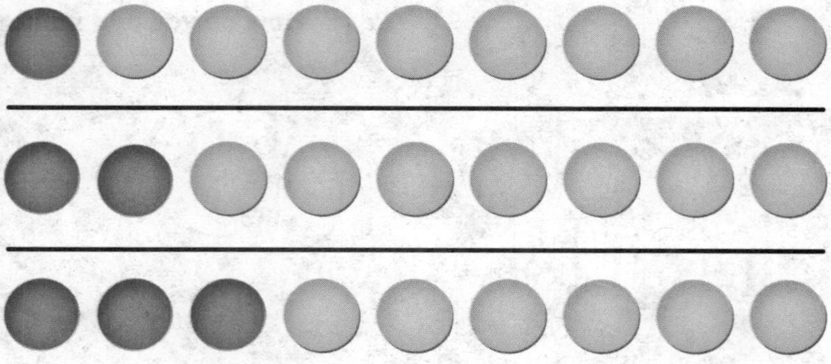

1 + 8 = 9

2 + 7 = 9

3 + 6 = 9

Use Work Mat 3 and  to show different ways to make a sum of 9. Write the numbers.

Ways to Make 9

1. ____ + ____ = 9	2. ____ + ____ = 9
3. ____ + ____ = 9	4. ____ + ____ = 9
5. ____ + ____ = 9	6. ____ + ____ = 9
7. ____ + ____ = 9	8. ____ + ____ = 9

Talk Math Why do you get the same sum when you add 6 + 3 and 7 + 2?

Name ______________________________

Your turn!

On My Own

Use Work Mat 3 and counters. Add.

9. $5 + 4 =$ _____

10. $2 + 5 =$ _____

11. $4 + 2 =$ _____

12. $5 + 3 =$ _____

13. $2 + 7 =$ _____

14. $4 + 3 =$ _____

15. $3 + 6 =$ _____

16. $4 + 4 =$ _____

17. $8 + 1 =$ _____

18. $6 + 1 =$ _____

19. $1 + 4 =$ _____

20. $2 + 6 =$ _____

21. $\begin{array}{r} 7 \\ +\ 1 \\ \hline \end{array}$

22. $\begin{array}{r} 0 \\ +\ 9 \\ \hline \end{array}$

23. $\begin{array}{r} 1 \\ +\ 8 \\ \hline \end{array}$

24. $\begin{array}{r} 4 \\ +\ 2 \\ \hline \end{array}$

25. $\begin{array}{r} 9 \\ +\ 0 \\ \hline \end{array}$

26. $\begin{array}{r} 6 \\ +\ 3 \\ \hline \end{array}$

Problem Solving

Processes & Practices

Write an addition number sentence.

27. There are 3 yellow fish and 6 red fish in a pond. How many fish are there in all?

_____ + _____ = _____ fish

28. There are 4 turtles in a pond. 5 turtles are walking to the pond. How many turtles are there in all?

_____ + _____ = _____ turtles

Write Math Is there more than one way to make 9? Explain.

Name ____________________

My Homework

Lesson 10
Ways to Make 9

Homework Helper

eHelp Need help? connectED.mcgraw-hill.com

There are many ways to make a sum of 9.

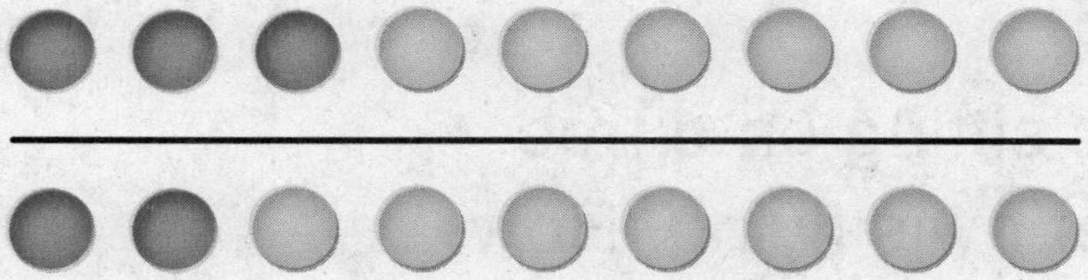

3 + 6 = 9

2 + 7 = 9

Practice

Write different ways to make 9.

Ways to Make 9

1. ______ + ______ = 9

2. ______ + ______ = 9

3. ______ + ______ = 9

4. ______ + ______ = 9

5. ______ + ______ = 9

6. ______ + ______ = 9

Add.

7. 4 + 5 = ______

8. 2 + 7 = ______

Add.

9. $\begin{array}{r} 6 \\ +\ 3 \\ \hline \end{array}$	10. $\begin{array}{r} 5 \\ +\ 1 \\ \hline \end{array}$	11. $\begin{array}{r} 9 \\ +\ 0 \\ \hline \end{array}$
12. $\begin{array}{r} 4 \\ +\ 3 \\ \hline \end{array}$	13. $\begin{array}{r} 7 \\ +\ 2 \\ \hline \end{array}$	14. $\begin{array}{r} 5 \\ +\ 3 \\ \hline \end{array}$

15. There are 5 owls sitting on a tree. There are 4 other owls flying. How many owls are there in all?

_____ owls

Test Practice

16. Which is not a way to make a sum of 9?

5 + 4	1 + 8	8 + 0	6 + 3
○	○	○	○

Math at Home Give your child 9 objects. Have your child show different ways to make two groups to show 9.

Name ______________________________

Ways to Make 10

Lesson 11

ESSENTIAL QUESTION
How do you add numbers?

Explore and Explain

Watch Tools

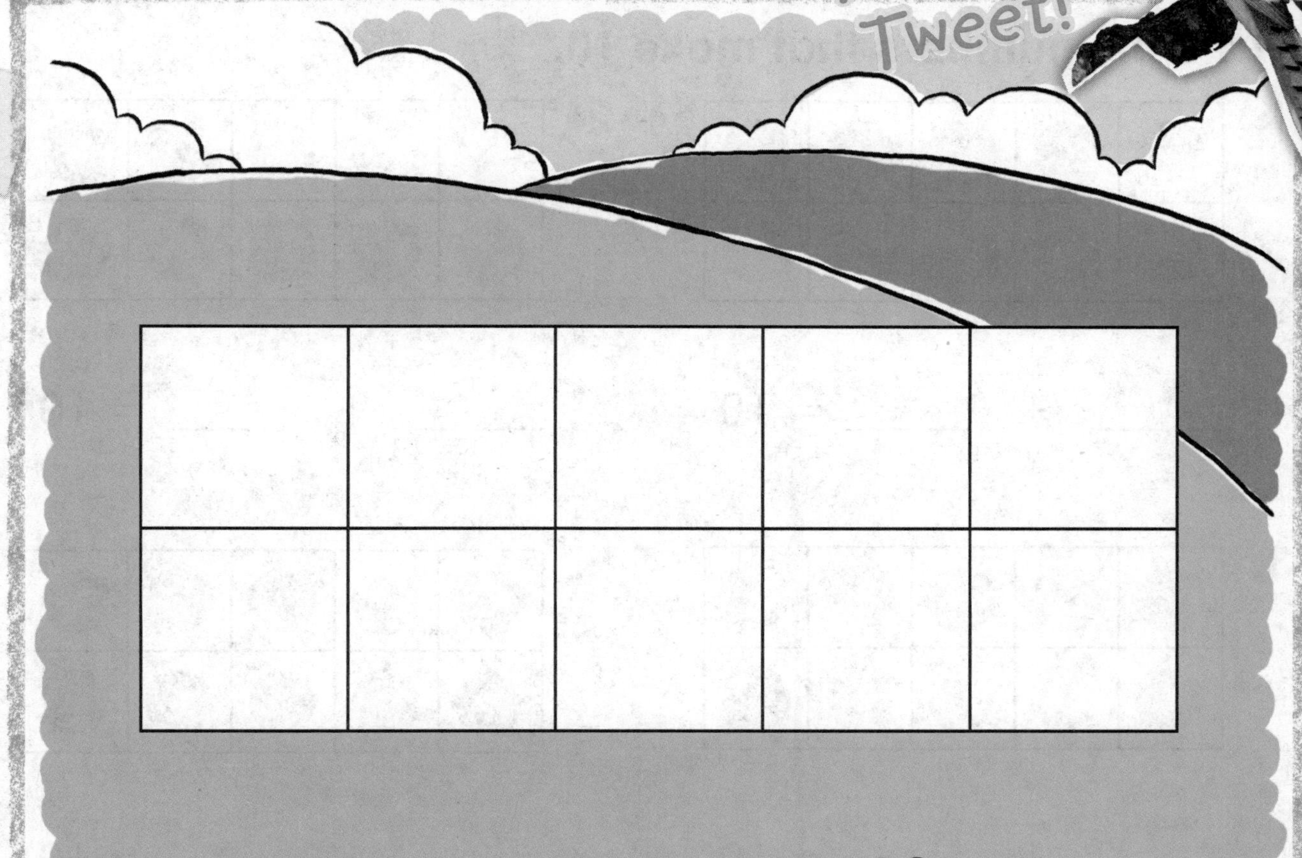

_____ + _____ = 10

Teacher Directions: Use counters to model. Find different ways to make 10. Trace and color counters to show one of the ways. Write the numbers.

See and Show

Processes & Practices

There are many ways to make 10.

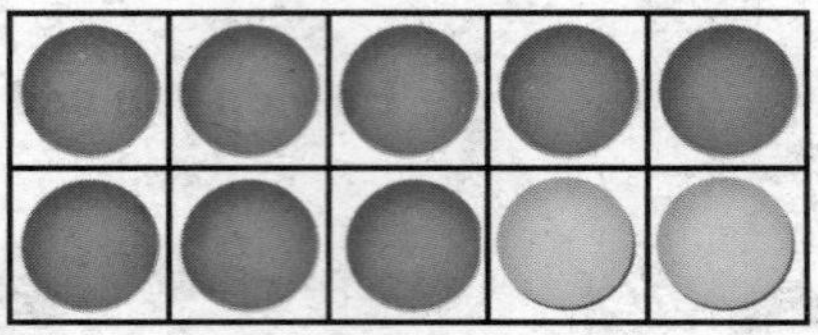

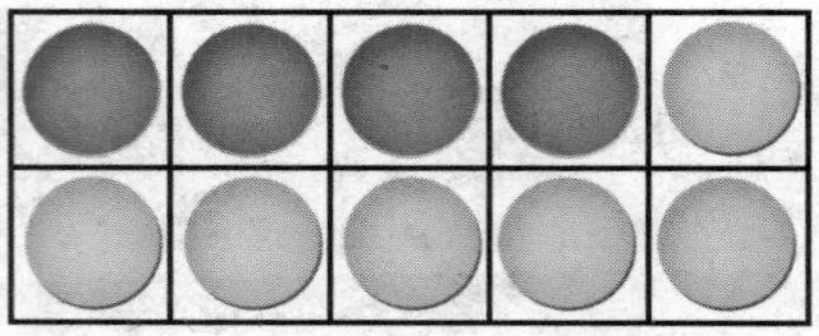

8 + 2 = 10   4 + 6 = 10

Write the numbers that make 10.

1.

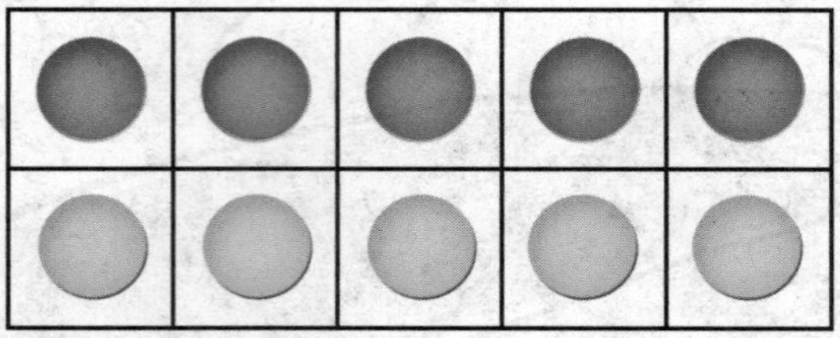

____ + ____ = 10

2.

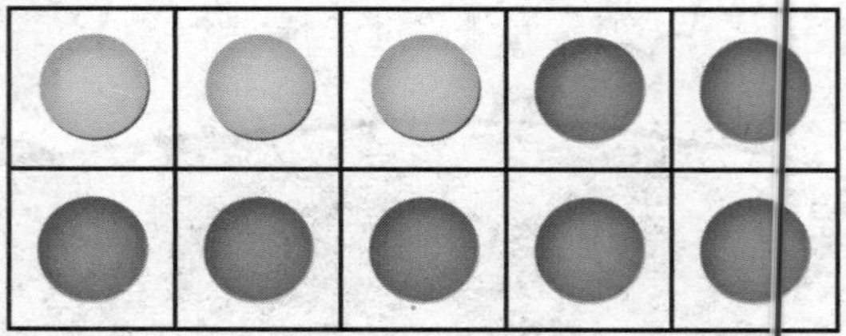

____ + ____ = 10

3.

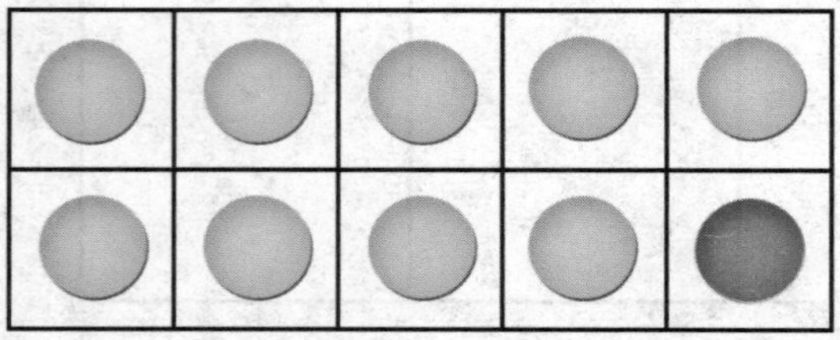

____ + ____ = 10

4.

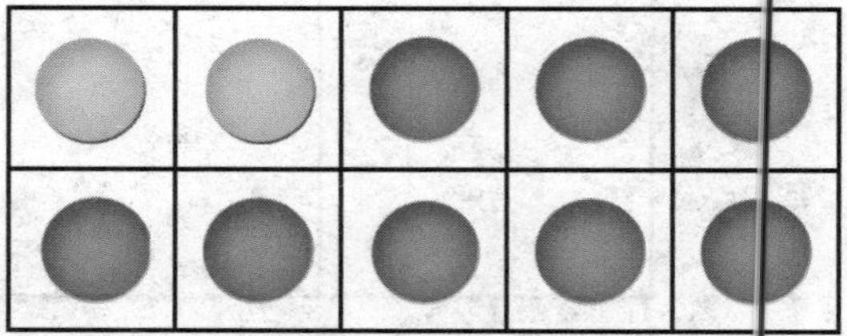

____ + ____ = 10

Talk Math Name all the ways to make 10 on a ten-frame using 2 numbers.

Name ....

On My Own

Write the numbers that make 10.

5.

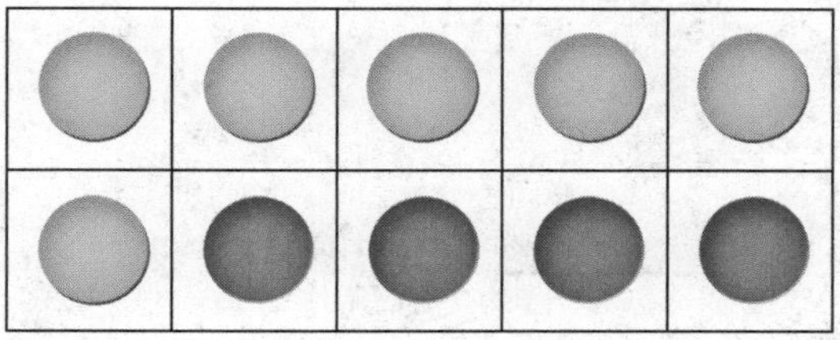

_____ + _____ = 10

6.

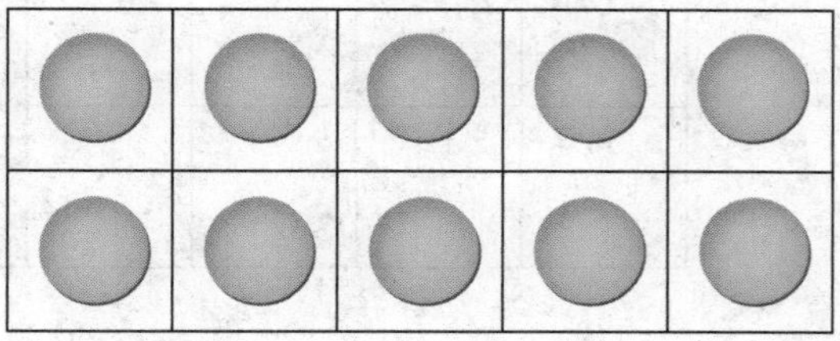

_____ + _____ = 10

7.

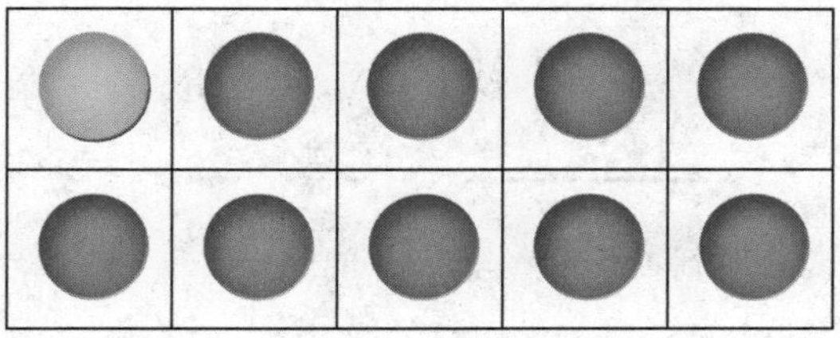

_____ + _____ = 10

8.

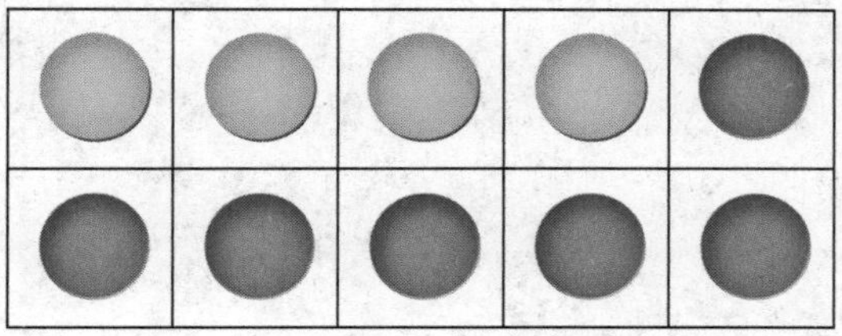

_____ + _____ = 10

Draw and color a way to make 10 using two numbers. Write the numbers.

9.

_____ + _____ = 10

10.

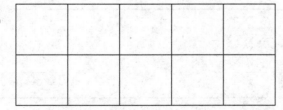

_____ + _____ = 10

Problem Solving

Processes &Practices

11. There are 5 red counters. How many yellow counters will make 10 in this ten-frame? Draw and color the counters.

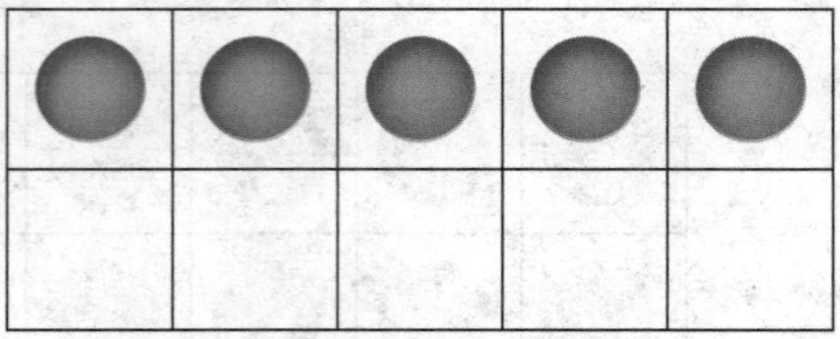

_____ counters

12. Joe saw 2 geese. How many more geese does he need to see so that he sees 10 geese in all?

_____ geese

HOT Problem Riley wrote this on the board. Tell why Riley is wrong. Make it right.

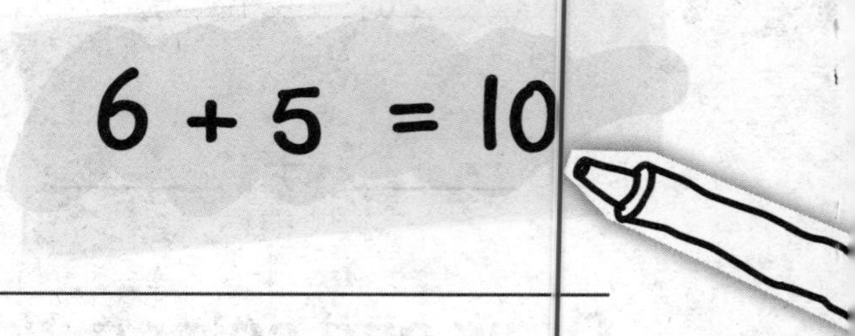

Name ______________________________

My Homework

Lesson 11

Ways to Make 10

Homework Helper

eHelp Need help? connectED.mcgraw-hill.com

There are many ways to make 10.

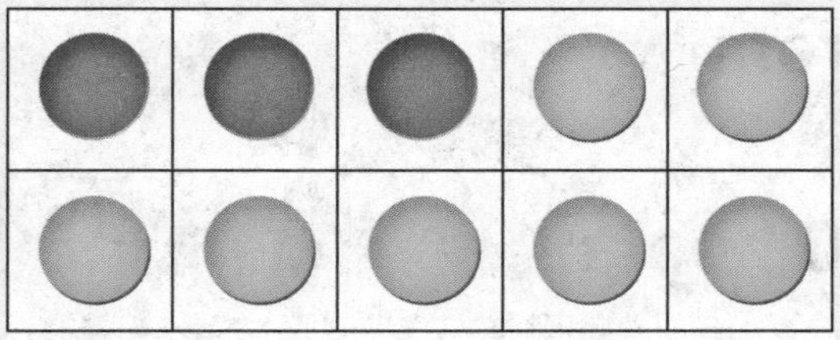

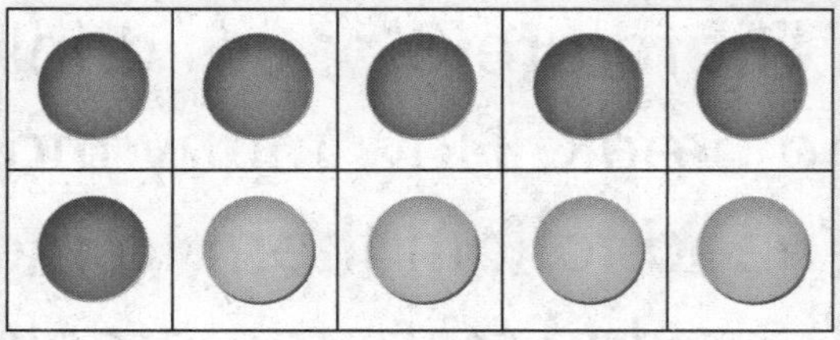

3 + 7 = 10 6 + 4 = 10

Practice

Write the numbers that make 10.

1.

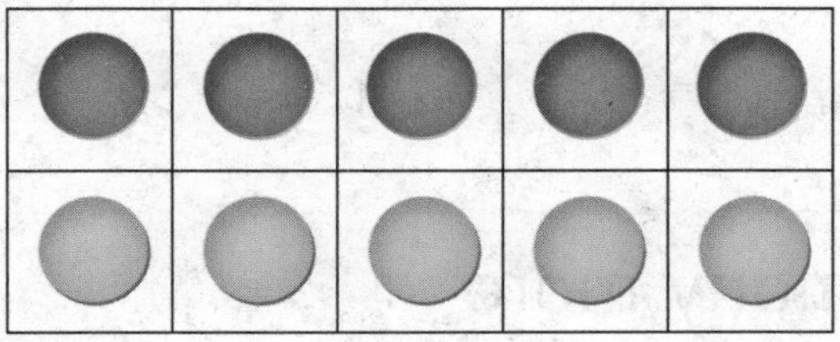

_____ + _____ = 10

2.

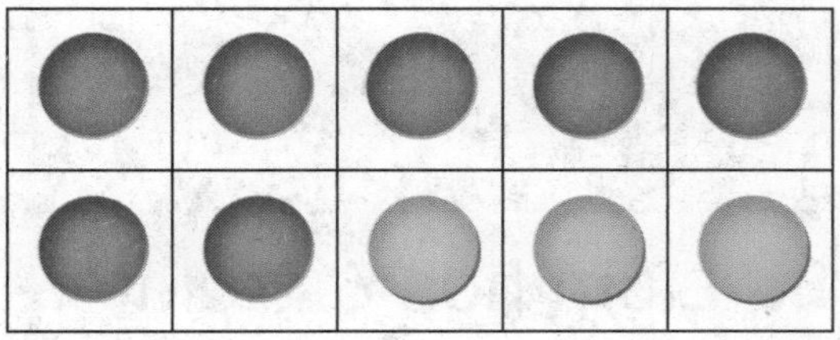

_____ + _____ = 10

3.

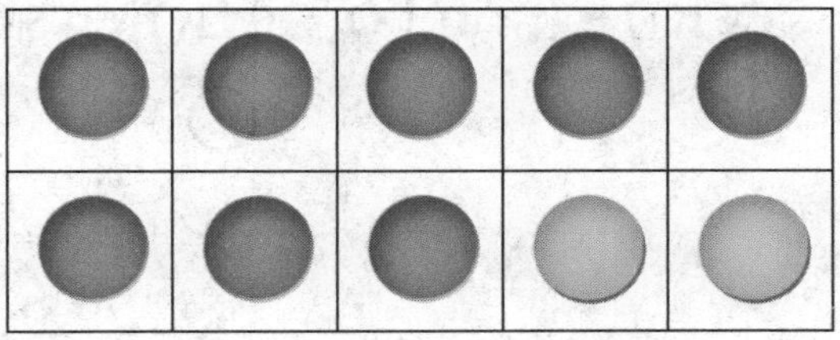

_____ + _____ = 10

4.

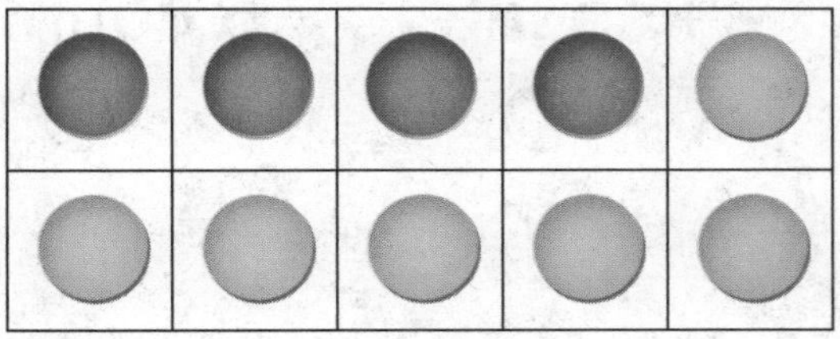

_____ + _____ = 10

Draw and color a way to make 10 using two numbers. Write the numbers.

5.

_____ + _____ = 10

6.

_____ + _____ = 10

7. There are 3 bears drinking from a creek. How many more bears need to join them to make 10 bears drinking from the creek?

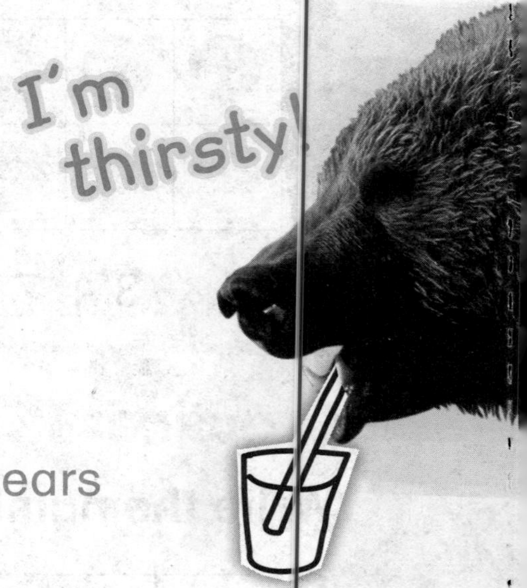

_____ bears

Test Practice

8. Liam has 6 counters. How many more counters does he need to have to make 10 counters?

○ 4 counters ○ 5 counters ○ 6 counters ○ 16 counters

Math at Home Give your child ten crayons. Have him or her put the crayons in two groups showing a way to make 10. Ask your child to show another way to make 10.

Name ______________________

Find Missing Parts of 10

Lesson 12

ESSENTIAL QUESTION
How do you add numbers?

Explore and Explain

Watch Tools

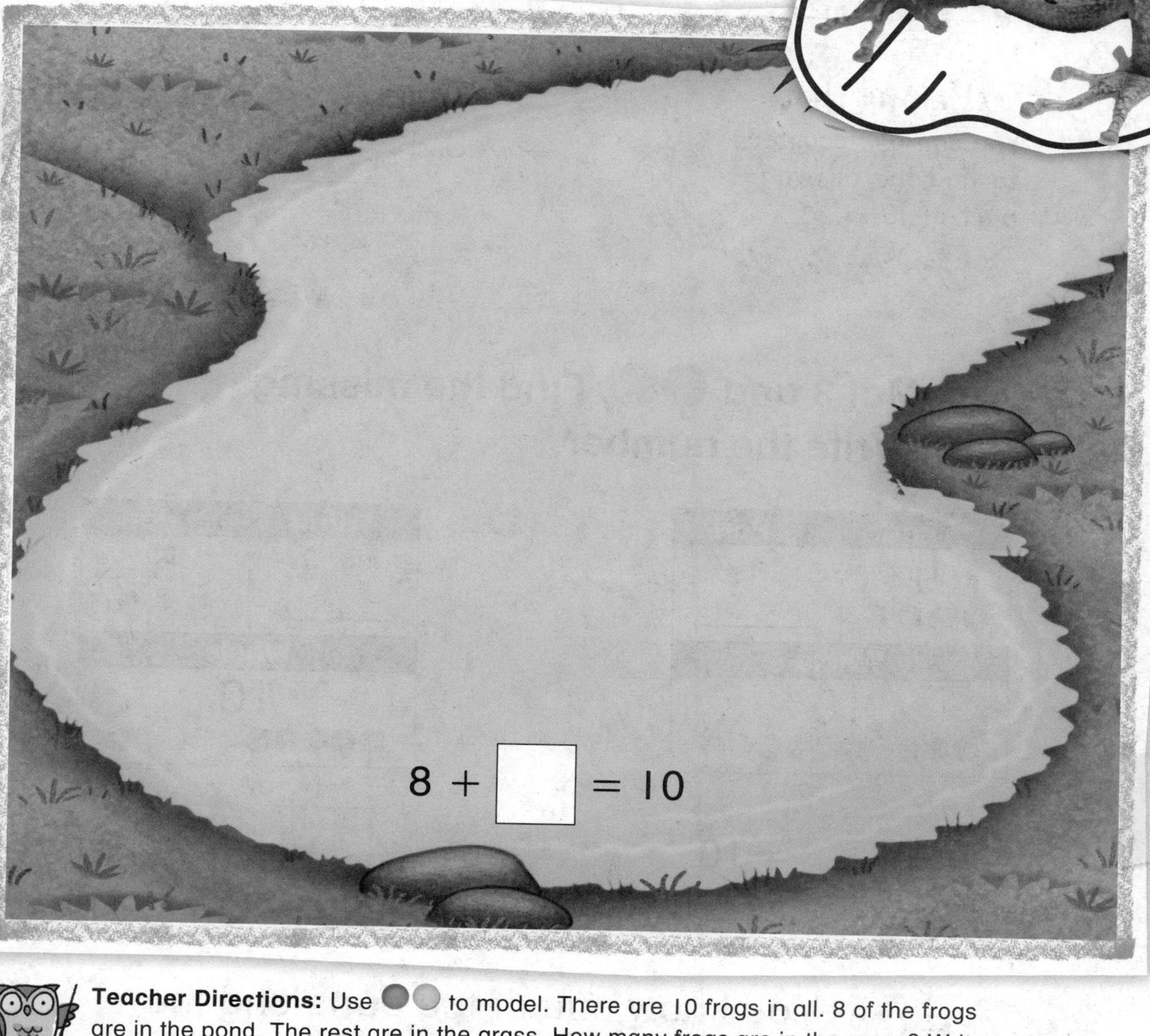

$8 + \square = 10$

Teacher Directions: Use ●● to model. There are 10 frogs in all. 8 of the frogs are in the pond. The rest are in the grass. How many frogs are in the grass? Write the missing part.

See and Show

Processes & Practices

The whole is 10. One part is 3.
What is the other part?

Part	Part
	___
Whole	

Part	Part
3	___
Whole	
10	

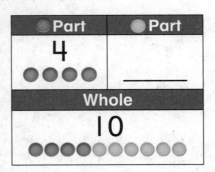

Helpful Hint
You can use counters to find the missing part of 10.

3 + [7] = 10

Use Work Mat 3 and counters. Find the missing part of 10. Write the number.

1.

Part	Part
4	___
Whole	
10	

4 + [ ] = 10

2.

Part	Part
___	5
Whole	
10	

[ ] + 5 = 10

Talk Math You know one of the parts and the whole, how do you find the other part?

Name ....

On My Own

Use Work Mat 3 and ●●. Find the missing part of 10. Write the number.

3.

Part	Part
7	___
Whole	
10	

7 + ☐ = 10

4.

Part	Part
___	5
Whole	
10	

☐ + 5 = 10

5.

Part	Part
___	8
Whole	
10	

☐ + 8 = 10

6.

Part	Part
1	___
Whole	
10	

1 + ☐ = 10

7.

Part	Part
3	___
Whole	
10	

3 + ☐ = 10

8.

Part	Part
___	6
Whole	
10	

☐ + 6 = 10

Problem Solving

Processes &Practices

9. There are 10 apples. 3 of the apples are green. The rest of the apples are red. How many apples are red?

______ apples

10. Javier has 10 leaves. 2 of the leaves are orange. The rest are yellow. How many yellow leaves does he have?

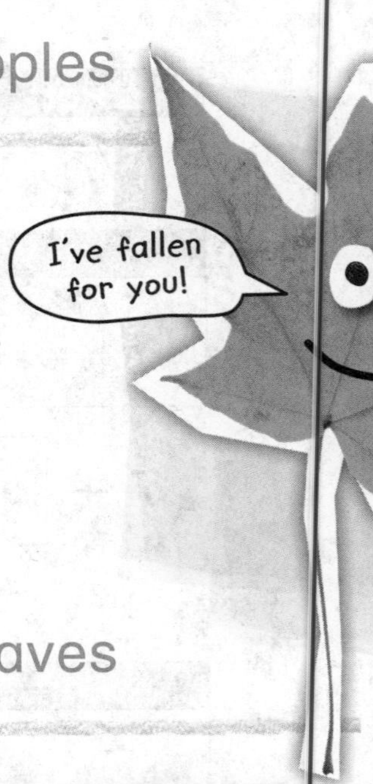

______ leaves

HOT Problem Angelo wrote the missing part. Tell why Angelo is wrong. Make it right.

Part	Part
2	7
Whole	
10	

Name ______________________

My Homework

Lesson 12
Find Missing Parts of 10

Homework Helper

eHelp Need help? connectED.mcgraw-hill.com

You can find the missing part of 10.

Part	Part
6	____
Whole	
10	

$6 + 4 = 10$

Part	Part
8	____
Whole	
10	

$8 + 2 = 10$

Practice

Find the missing part of 10. Write the number.

1.

Part	Part
3	____
Whole	
10	

3 + ☐ = 10

2.

Part	Part
____	5
Whole	
10	

☐ + 5 = 10

Find the missing part of 10. Write the number.

3.

Part	Part
6	___
Whole	
10	

6 + ☐ = 10

4.

Part	Part
___	1
Whole	
10	

☐ + 1 = 10

5. Amaya sees 10 bugs on a log. 8 of them are black. The rest of the bugs are red. How many bugs are red?

_____ bugs

Test Practice

6. Carter sees 10 eagles. 5 of the eagles are flying. The rest of the eagles are in their nest. How many eagles are in their nest?

15 eagles ○ 9 eagles ○ 10 eagles ○ 5 eagles ○

Math at Home Create number cards with the numbers 0 to 10 on them. Show one number card. Have your child find the other number card that shows the number needed to make 10.

Name

True and False Statements

Lesson 13

ESSENTIAL QUESTION
How do you add numbers?

Explore and Explain

Tools

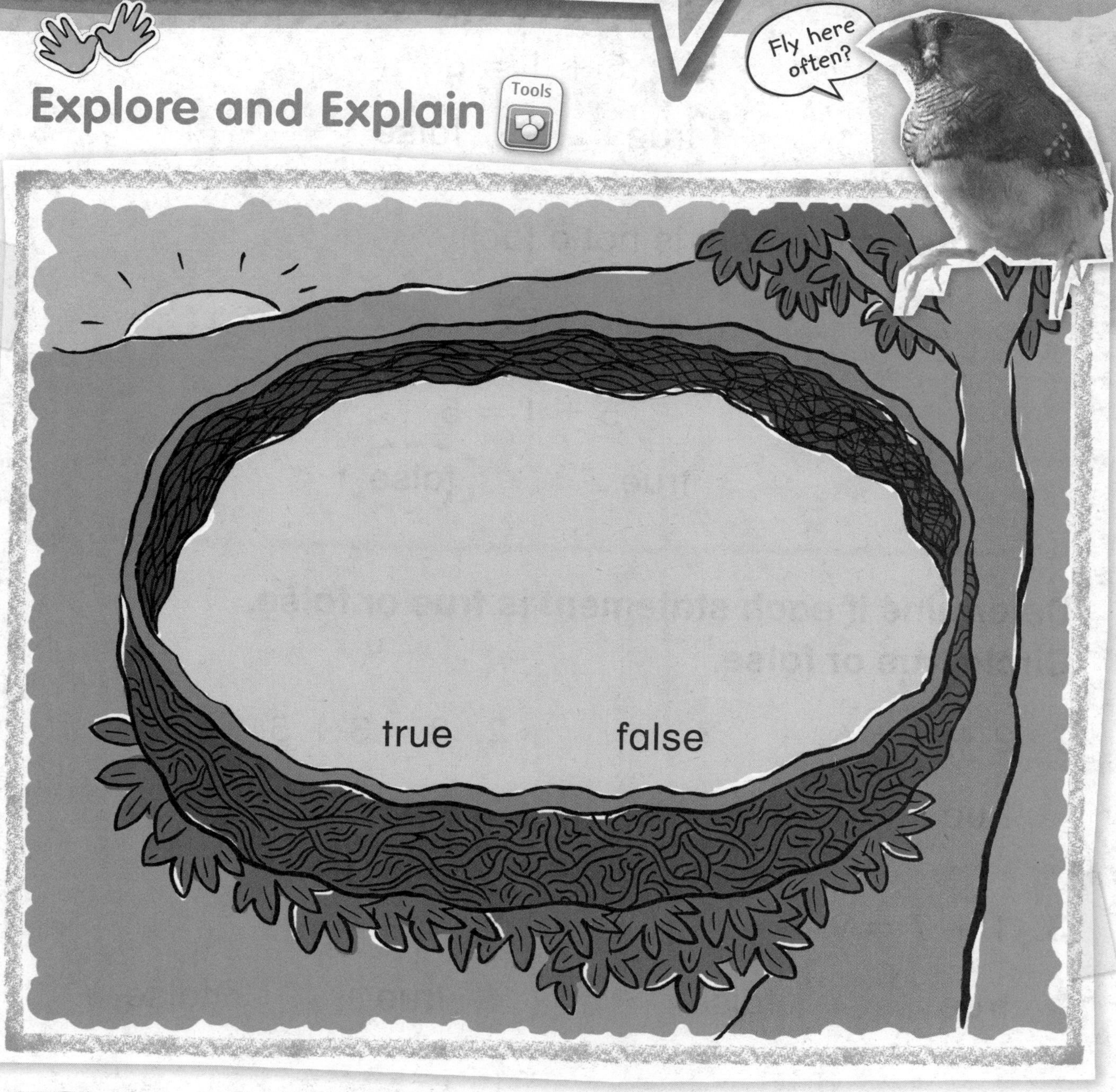
Fly here often?
true
false

Teacher Directions: Use [counters] to model. There are 3 yellow and 3 red eggs in a nest. There are 7 eggs in all. Is this true or false? Circle the word. Trace the counters you used to show the problem.

See and Show

Statements can be true or false.

A **true** statement is a fact.

5 + 1 = 6

true   false

A **false** statement is not a fact.

5 + 1 = 5

true   false

Determine if each statement is true or false.
Circle true or false.

1. 2 + 4 = 6

true   false

2. 8 = 3 + 5

true   false

3. 1 + 7 = 9

true   false

4. 7 = 7

true   false

Talk Math Tell your own false addition statement to a classmate.

Name ....

On My Own

Determine if each statement is true or false. Circle true or false.

5. $1 + 3 = 5$

true false

6. $5 + 5 = 10$

true false

7. $3 + 5 = 7$

true false

8. $9 = 9 + 0$

true false

9. $6 + 2 = 8$

true false

10. $5 + 2 = 4$

true false

11. $3 = 3$

true false

12. $4 + 2 = 7$

true false

13. $9 = 8 + 2$

true false

14. $2 + 5 = 7$

true false

15. $\begin{array}{r} 4 \\ +\ 4 \\ \hline 8 \end{array}$

true false

16. $\begin{array}{r} 6 \\ +\ 1 \\ \hline 5 \end{array}$

true false

Problem Solving

Processes & Practices

Determine if the word problem is true or false. Circle true or false.

17. There are 4 children bird watching. 3 more children join them. There are 7 children bird watching in all.

true   false

18. There are 3 mice playing in a field. 4 more mice join them. There are 6 mice in the field in all.

true   false

Write Math Is $6 + 2 = 3 + 6$ a true or false math statement? Explain.

Name ....

My Homework

Lesson 13
True and False Statements

Homework Helper

eHelp Need help? connectED.mcgraw-hill.com

Math statements are true or false.

5 + 1 = 4

true false

3 + 2 = 5

true false

Practice

Determine if each statement is true or false. Circle true or false.

1. 3 + 1 = 4

   true false

2. 0 = 4 + 0

   true false

3. 5 + 4 = 9

   true false

4. 10 = 6 + 4

   true false

5. 3 + 6 = 10

   true false

6. 4 + 1 = 5

   true false

Determine if each statement is true or false. Circle true or false.

7. 9 = 8 + 1

true false

8. 6 + 2 = 3 + 6

true false

9.
$$\begin{array}{r} 1 \\ +\ 0 \\ \hline 0 \end{array}$$

true false

10.
$$\begin{array}{r} 3 \\ +\ 3 \\ \hline 6 \end{array}$$

true false

11. There is 1 dog at the park.
4 more dogs come to the park.
There are 6 dogs at the park in all.

true false

Vocabulary Check

Vocab

Draw lines to match.

12. **true** Something that is not a fact.

13. **false** Something that is a fact.

Math at Home Tell your child a false addition number sentence. Ask your child if it is true or false. Have your child make it true.

Name ....

Processes & Practices

Fluency Practice

Add.

1. 4 + 6 = ______
2. 5 + 4 = ______
3. 3 + 2 = ______
4. 2 + 6 = ______
5. 2 + 5 = ______
6. 1 + 3 = ______
7. 7 + 1 = ______
8. 0 + 9 = ______
9. 1 + 1 = ______
10. 3 + 7 = ______
11. 4 + 4 = ______
12. 5 + 1 = ______
13. 7 + 3 = ______
14. 2 + 7 = ______
15. 4 + 3 = ______
16. 3 + 0 = ______
17. 0 + 5 = ______
18. 8 + 2 = ______
19. 5 + 3 = ______
20. 9 + 1 = ______
21. 4 + 5 = ______
22. 1 + 2 = ______
23. 7 + 0 = ______
24. 3 + 5 = ______

Fluency Practice

Add.

1. 7 + 2 = ____

2. 4 + 3 = ____

3. 6 + 2 = ____

4. 1 + 8 = ____

5. 2 + 4 = ____

6. 4 + 5 = ____

7. 2 + 2 = ____

8. 5 + 1 = ____

9. 4 + 6 = ____

10. 2 + 7 = ____

11. 3 + 3 = ____

12. 5 + 3 = ____

13. 6 + 1 = ____

14. 2 + 8 = ____

15. 0 + 6 = ____

16. 1 + 9 = ____

17. 0 + 4 = ____

18. 7 + 3 = ____

19. 3 + 5 = ____

20. 10 + 0 = ____

Name ________________________

My Review

Chapter 1

Addition Concepts

Vocabulary Check

Complete each sentence.

add **part** **sum** **whole**

1. Joining two parts together makes a ______________.

2. A ______________ is one of the groups that are joined when adding.

3. When you ______________ two numbers together, you come up with the sum.

4. The answer in an addition number sentence is called the ______________.

Concept Check

Write the addition number sentence.

5. 

____ ◯ ____ ◯ ____

6. 

____ ◯ ____ ◯ ____

Add.

7. 1 + 6 = ____

8. 8 + 0 = ____

9. 3 + 2 = ____

10. 4 + 4 = ____

11. $\begin{array}{r} 4 \\ +\ 0 \\ \hline \end{array}$

12. $\begin{array}{r} 3 \\ +\ 1 \\ \hline \end{array}$

13. $\begin{array}{r} 5 \\ +\ 4 \\ \hline \end{array}$

14. $\begin{array}{r} 2 \\ +\ 6 \\ \hline \end{array}$

Show a way to make the sum. Color the ○. Write the numbers.

15. ○○○○○○ ____ + ____ = 6

16. ○○○○○ ____ + ____ = 5

Find the missing part of ten. Write the number.

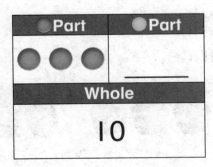

17.

Part	Part
● ● ●	____
Whole	
10	

3 + ☐ = 10

18.

Part	Part
____	●
Whole	
10	

☐ + 1 = 10

Name ....

Problem Solving

Write an addition number sentence.

19. Liliana finds 3 twigs. Enrique finds 2 twigs. How many twigs do they find in all?

_____ + _____ = _____ twigs

Determine if the statement is true or false. Circle true or false.

20. There are 3 bears yawning. 4 more bears begin yawning. There are 8 bears yawning in all.

true false

Test Practice

21. Miranda sees 6 stars in the sky. Madison sees 3 stars in the sky. How many stars do they see in all?

2 stars	3 stars	9 stars	10 stars
○	○	○	○

Reflect

Chapter 1
Answering the Essential Question

Show the ways to answer.

Find the whole.

Part	Part
3	5
Whole	
___	

Write an addition number sentence.

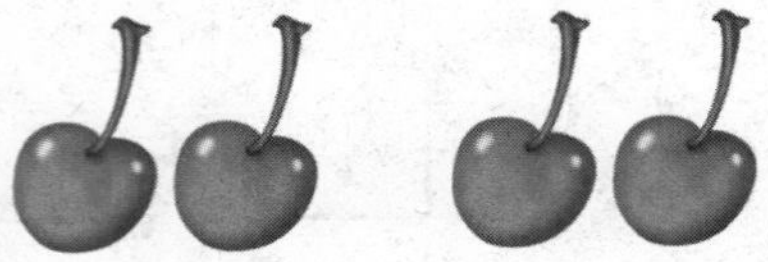

___ + ___ = ___

ESSENTIAL QUESTION
How do you add numbers?

Add.

$$\begin{array}{r} 3 \\ +\ 4 \\ \hline \square \end{array}$$

Add zero.

0 + 9 = ___

Now I Know!

Chapter

2 Subtraction Concepts

ESSENTIAL QUESTION

How do you subtract numbers?

Let's Go on a Safari!

Watch a video!

Watch

Name ....

Chapter 2 Project

Subtraction Storybook

1. Decide how to design the cover of your subtraction storybook.
2. Write a title for your book.
3. Use drawings and words to design your cover for the book below.

Name ______________________________

Am I Ready?

Write how many.

1. 

______

2. 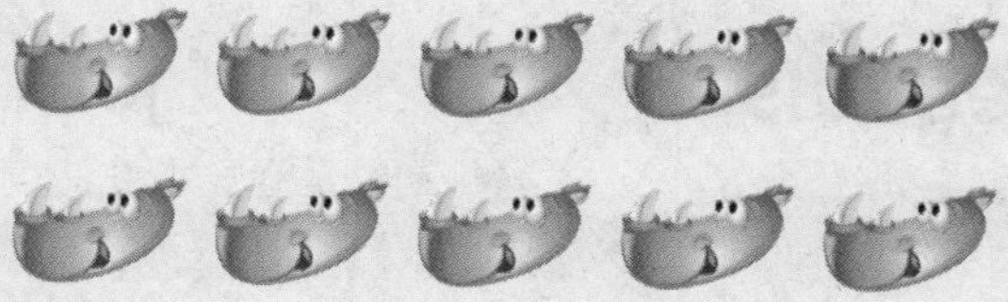

______

Draw circles to show each number.

3. 6

4. 4

Put an X on 2 frogs. Write how many are left.

5. 

______ frogs

How Did I Do?

Shade the boxes to show the problems you answered correctly.

1	2	3	4	5

Name

My Math Words

Vocab

Review Vocabulary

equals   join   take away

Trace the words. Then draw a picture in the box to show what the word means.

Word	My Example
equals	
join	
take away	

My Vocabulary Cards

Vocab abc

Processes & Practices

Lesson 2-7

compare

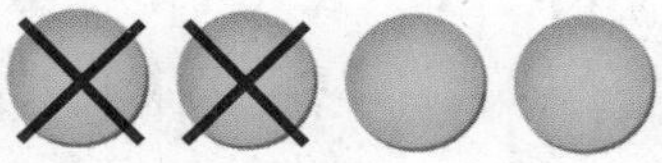

2 more

2 fewer

Lesson 2-3

difference

4 − 1 = 3

Lesson 2-3

minus (−)

6 − 2 = 4

Lesson 2-13

related facts

1 + 2 = 3　　3 − 1 = 2

2 + 1 = 3　　3 − 2 = 1

Lesson 2-2

subtract

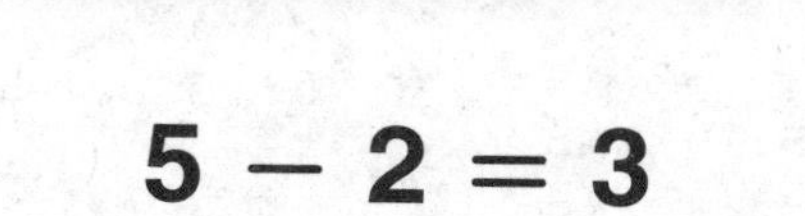

5 − 2 = 3

Lesson 2-3

subtraction number sentence

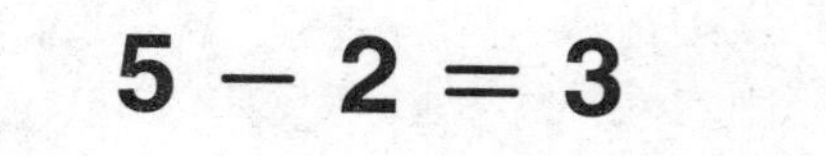

5 − 2 = 3

Teacher Directions: Ideas for Use

- Have students draw different examples for each card.
- Have students arrange cards to show words with similar meanings. Have them explain the meaning of their grouping.

The answer to a subtraction problem.	Look at groups of objects, shapes, or numbers and see how they are alike or different.
Basic facts using the same numbers.	The sign used to show subtraction.
An expression using numbers and the − and = signs.	To take away.

My Vocabulary Cards

Vocab

abc

Processes
&Practices

Teacher Directions:
More Ideas for Use

- Use the blank cards to write your own vocabulary words.
- Have students draw pictures on the blank cards to show the meaning of each new vocabulary word.

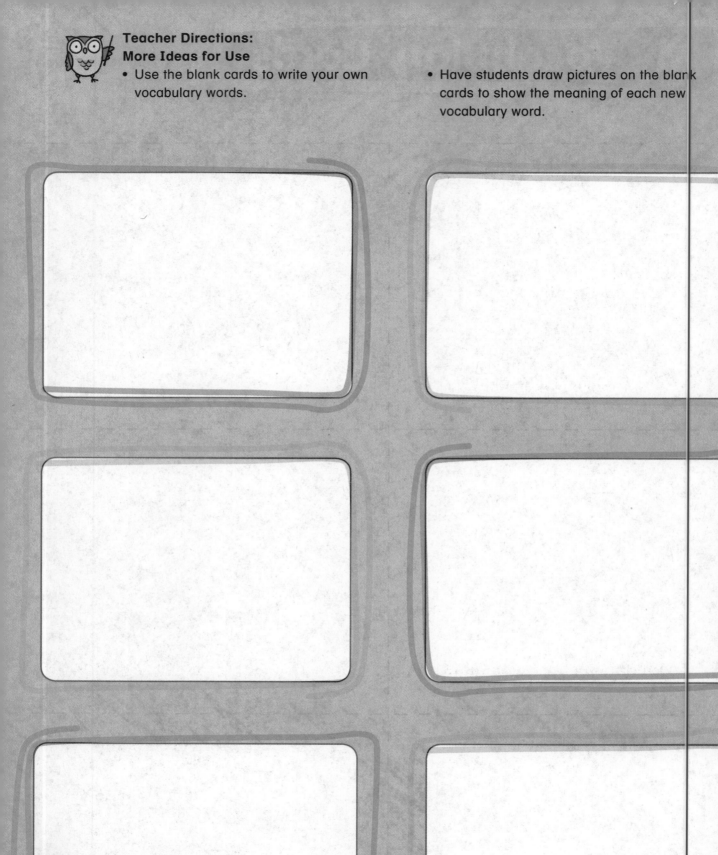

My Foldable

FOLDABLES® Follow the steps on the back to make your Foldable.

4 − 0

4 − 1

4 − 2

4 − 3

4 − 4

FOLDABLES®
Study Organizer

1
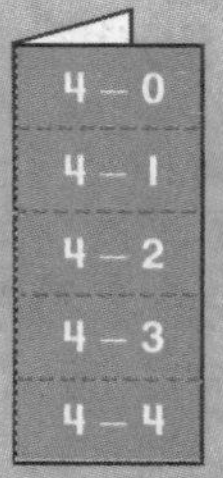

2
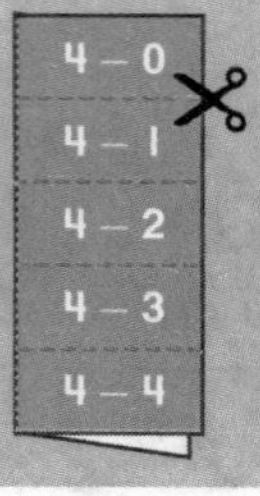

3
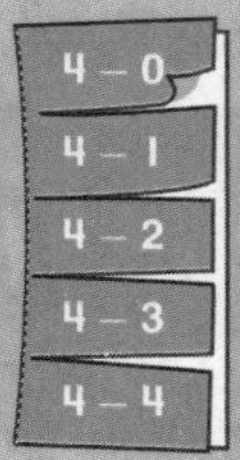

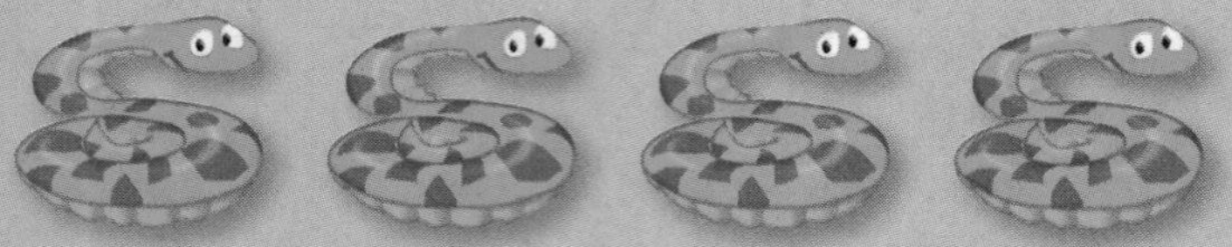

Name ....

Subtraction Stories

Lesson 1

ESSENTIAL QUESTION
How do you subtract numbers?

Explore and Explain

Watch Tools

______ dragonflies

Teacher Directions: Use ●● to model. There are 7 dragonflies sitting on a flower. 2 fly away. How many dragonflies are left on the flower? Write the number.

See and Show

Processes
&Practices

There are 6 cats on a fence.
1 cat jumps down off of the fence.

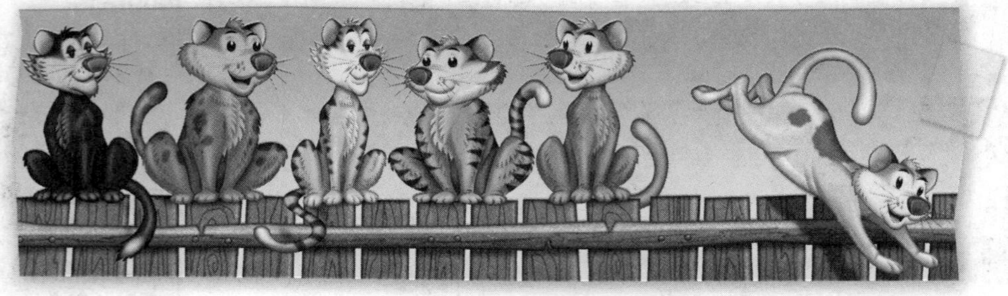

How many cats are left on the fence? 5 cats

Tell a number story. Use ●●. Write how many are left.

1.

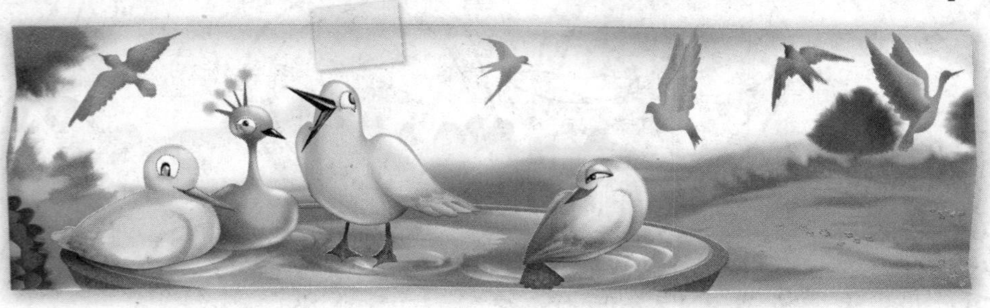

How many birds are left in the bird bath? _______ birds

2.

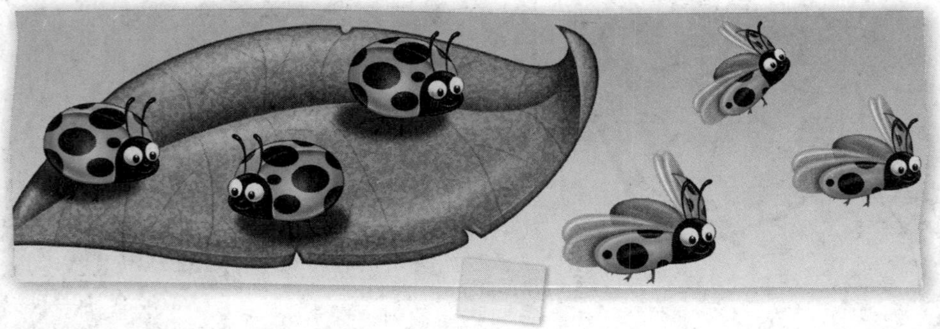

How many ladybugs are left on the leaf? _______ ladybugs

Talk Math How are addition and subtraction stories different?

Name ______________________

On My Own

Tell a number story. Use counters. Write how many are left.

3.

How many ants are left on the leaf? ______ ants

4.

How many butterflies are left on the bush?

______ butterflies

5.

How many kites are left on the ground? ______ kites

Problem Solving

Processes
&Practices

Use to solve.

6. There are 4 people hiking on a trail. 2 of the people go home. How many people are left?

______ people

7. There are 8 bees near the hive. 3 bees fly away. How many bees are left?

______ bees

HOT Problem There are 6 tigers sleeping under a tree. 2 tigers wake up. The answer is 4 tigers. What is the question?

Name ....

My Homework

Lesson 1
Subtraction Stories

Homework Helper

eHelp Need help? connectED.mcgraw-hill.com

There are 4 rabbits playing together.
2 rabbits hop away.

How many rabbits are left playing near the carrots?

2 rabbits

Practice

Tell a number story. Use pennies to model if needed. Write how many are left.

1.

How many birds are left on the window? ______ birds

Tell a number story. Use pennies to model if needed. Write how many are left.

2.

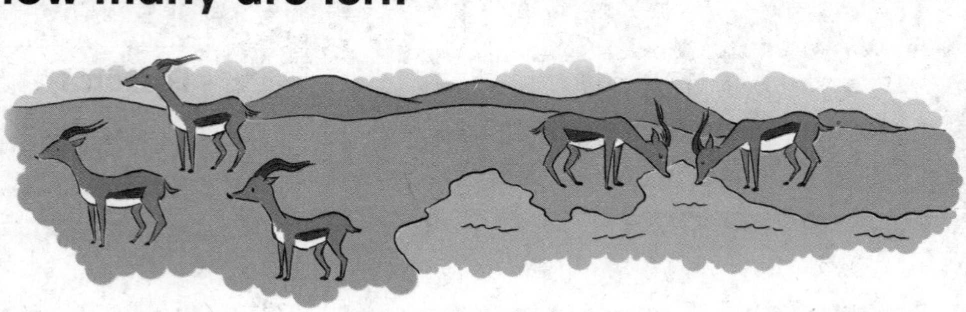

How many animals are still drinking water?

_____ animals

3.

How many animals are left standing still?

_____ animals

Test Practice

4. There are 8 hippos in a pond. 5 of the hippos get out. How many hippos are left in the pond?

9	6	3	0
○	○	○	○

Math at Home Tell subtraction stories to your child. Have your child use objects such as stuffed animals, toy cars, or crayons to model each story.

Name ....

Model Subtraction

Lesson 2

ESSENTIAL QUESTION
How do you subtract numbers?

Which toy is your favorite?

Explore and Explain

Watch Tools

Part	Part
8	______
Whole	
10	

Teacher Directions: Use ●● to model. There are 10 toys in a toy box. Marty takes 8 toys out of the toy box. How many toys are left? Write the number.

See and Show

Processes & Practices

When you know the whole and one part, you can **subtract** to find the other part.

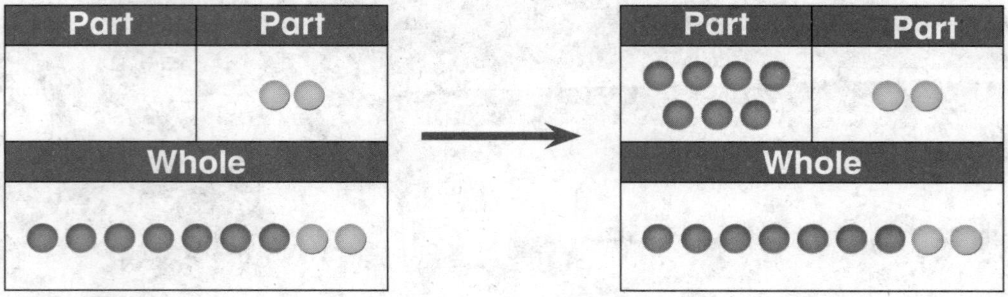

Part	Part
___	2
Whole	
9	

→

Part	Part
7	2
Whole	
9	

Use Work Mat 3 and counters to subtract.

1.

Part	Part
4	___
Whole	
5	

2.

Part	Part
___	6
Whole	
8	

Talk Math You have 10 counters. 3 are yellow. Tell how you would use the part-part-whole mat to find how many are red. Explain.

Part	Part
	___
Whole	

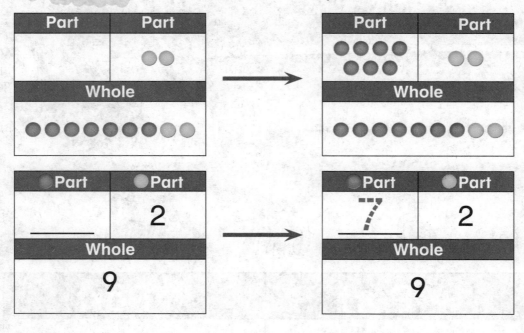

Name ....

On My Own

Use Work Mat 3 and counters to subtract.

3.

Part	Part
3	___
Whole	
8	

4.

Part	Part
___	3
Whole	
7	

5.

Part	Part
4	___
Whole	
9	

6.

Part	Part
___	1
Whole	
4	

7.

Part	Part
4	___
Whole	
6	

8.

Part	Part
___	3
Whole	
5	

9.

Part	Part
___	5
Whole	
10	

10.

Part	Part
1	___
Whole	
7	

Real World Problem Solving

Processes & Practices

Solve. Use Work Mat 3 and [counters] if needed.

11. Clara saw 6 eagles on a limb. 2 eagles flew away. How many eagles were left on the limb?

_____ eagles

12. There are 10 crocodiles in a pond. 3 get out of the pond. How many crocodiles are left in the pond?

_____ crocodiles

Write Math The whole is 10 and one of the parts is 10. What is the other part? Explain.

Name ....

My Homework

Lesson 2
Model Subtraction

Homework Helper

eHelp

Need help? connectED.mcgraw-hill.com

When you know the whole and one of the parts, you can subtract to find the other part.

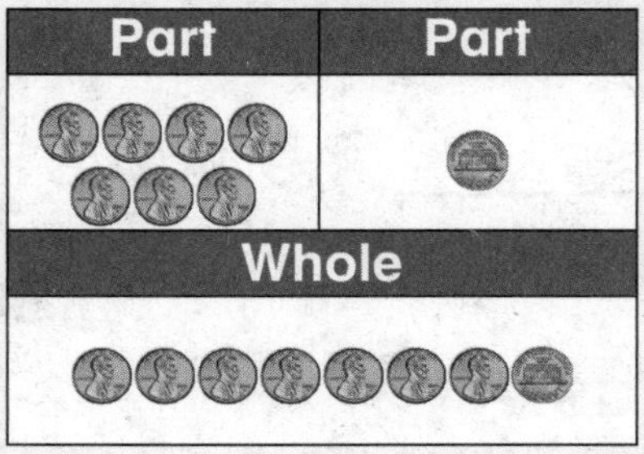

Part	Part
7	1
Whole	
8	

Practice

Use pennies to subtract. Write the number.

1.

Part	Part
1	____
Whole	
5	

2.

Part	Part
2	____
Whole	
10	

3.

Part	Part
1	____
Whole	
6	

4.

Part	Part
6	____
Whole	
9	

Use pennies to subtract. Write the number.

5.

Part	Part
6	____
Whole	
7	

6.

Part	Part
4	____
Whole	
8	

7. There are 7 monkeys hanging from a branch. 3 monkeys go away. How many monkeys are still on the branch?

____ monkeys

8. There are 9 apes eating bananas. 1 ape stops eating. How many apes are still eating bananas?

____ apes

Vocabulary Check

Circle the correct answer.

sum **subtract**

9. You know the whole and one part. You can ____________ to find the other part.

Math at Home Have your child use small objects such as cereal or beans to model subtraction.

Name

Subtraction Number Sentences

Lesson 3

ESSENTIAL QUESTION
How do you subtract numbers?

Explore and Explain

Watch Tools

Time Out!
___ − ___ = ___
Write your subtraction sentence here.

Teacher Directions: Use ● ● to model. There are 7 zebras playing in a field. 5 of the zebras get in the pond. Draw an X on the zebras that get in the pond. How many zebras are still playing in the field? Write the subtraction number sentence.

See and Show

Processes & Practices

You can write a subtraction number sentence.

See

Say 5 **minus** 2 equals 3.

Write 5 – 2 = 3

5 – 2 = 3 is a **subtraction number sentence.**

3 is the **difference.**

Write a subtraction number sentence.

1.

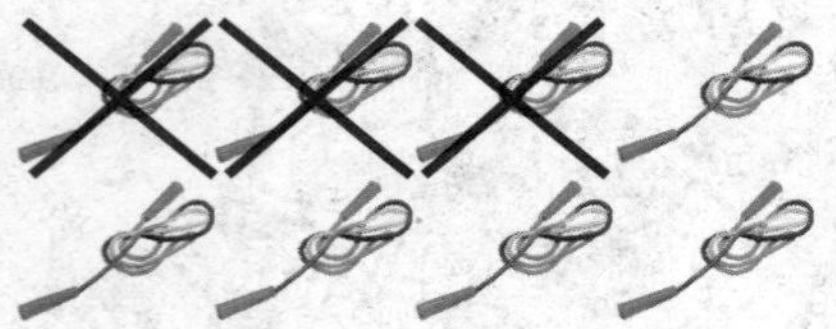

___ ○ ___ ○ ___

2.

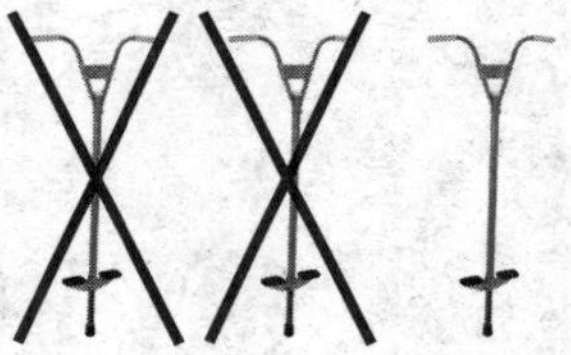

___ ○ ___ ○ ___

3.

___ ○ ___ ○ ___

4.

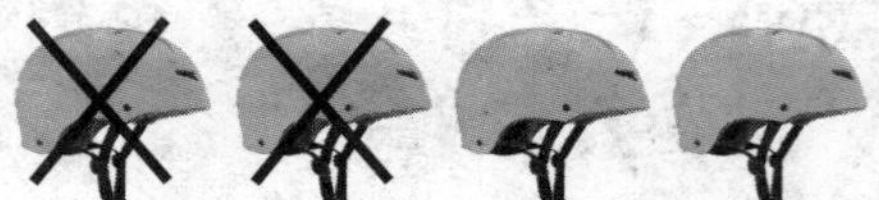

___ ○ ___ ○ ___

Talk Math What does – mean?

Name ____________________

On My Own

Write a subtraction number sentence.

5.

___ ◯ ___ ◯ ___

6.

___ ◯ ___ ◯ ___

7.

___ ◯ ___ ◯ ___

8.

___ ◯ ___ ◯ ___

9.

___ ◯ ___ ◯ ___

10.
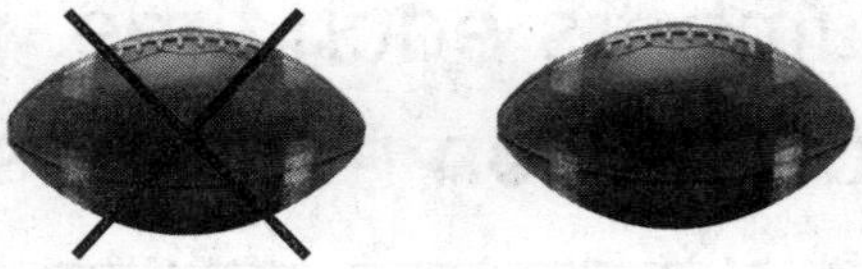

___ ◯ ___ ◯ ___

11.

___ ◯ ___ ◯ ___

12.

___ ◯ ___ ◯ ___

Problem Solving

Processes & Practices

13. There are 5 cars racing. 2 cars stop.
How many cars are still racing?

____ ◯ ____ ◯ ____ cars

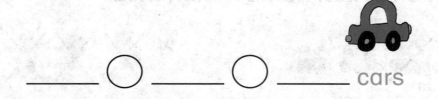

14. Kayla has 4 rockets. She gives some rockets away. She has 1 rocket left.
How many rockets did she give away?

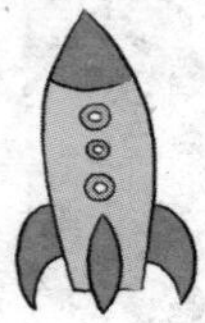

____ ◯ ____ ◯ ____ rockets

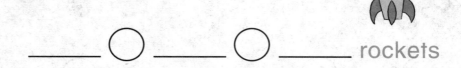

HOT Problem By using two of these numbers each time, write as many subtraction number sentences as you can.

2 8 6

Name ....

My Homework

Lesson 3

Subtraction Number Sentences

Homework Helper

eHelp **Need help?** connectED.mcgraw-hill.com

You can write a subtraction number sentence.

8 − 2 = 6

Practice

Write a subtraction number sentence.

1.

_____ ◯ _____ ◯ _____

2.

_____ ◯ _____ ◯ _____

3.

_____ ◯ _____ ◯ _____

4.

_____ ◯ _____ ◯ _____

Write a subtraction number sentence.

5. 

____ ◯ ____ ◯ ____

6. 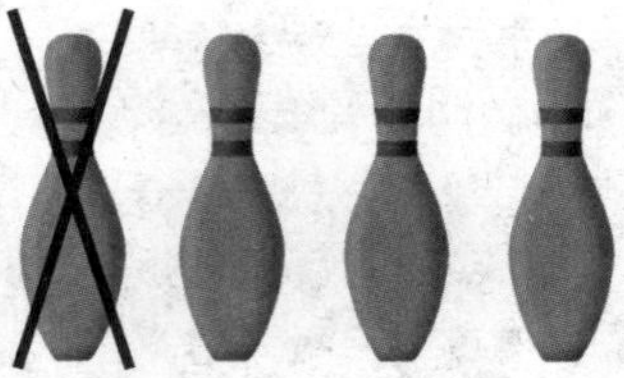

____ ◯ ____ ◯ ____

7. There are 7 elephants in a pond. 3 elephants get out. How many elephants are left?

____ ◯ ____ ◯ ____ elephants

Vocabulary Check

Complete each sentence.

difference **subtraction number sentence**

8. 6 − 4 = 2 is a ______________________

______________.

9. In 3 − 2 = 1, the ______________________ is 1.

Math at Home Using buttons, beans, or cereal, have your child act out subtraction stories. Have him or her write subtraction number sentences for the stories.

Name ____________________

Subtract 0 and All

Lesson 4

ESSENTIAL QUESTION
How do you subtract numbers?

Explore and Explain

Watch Tools

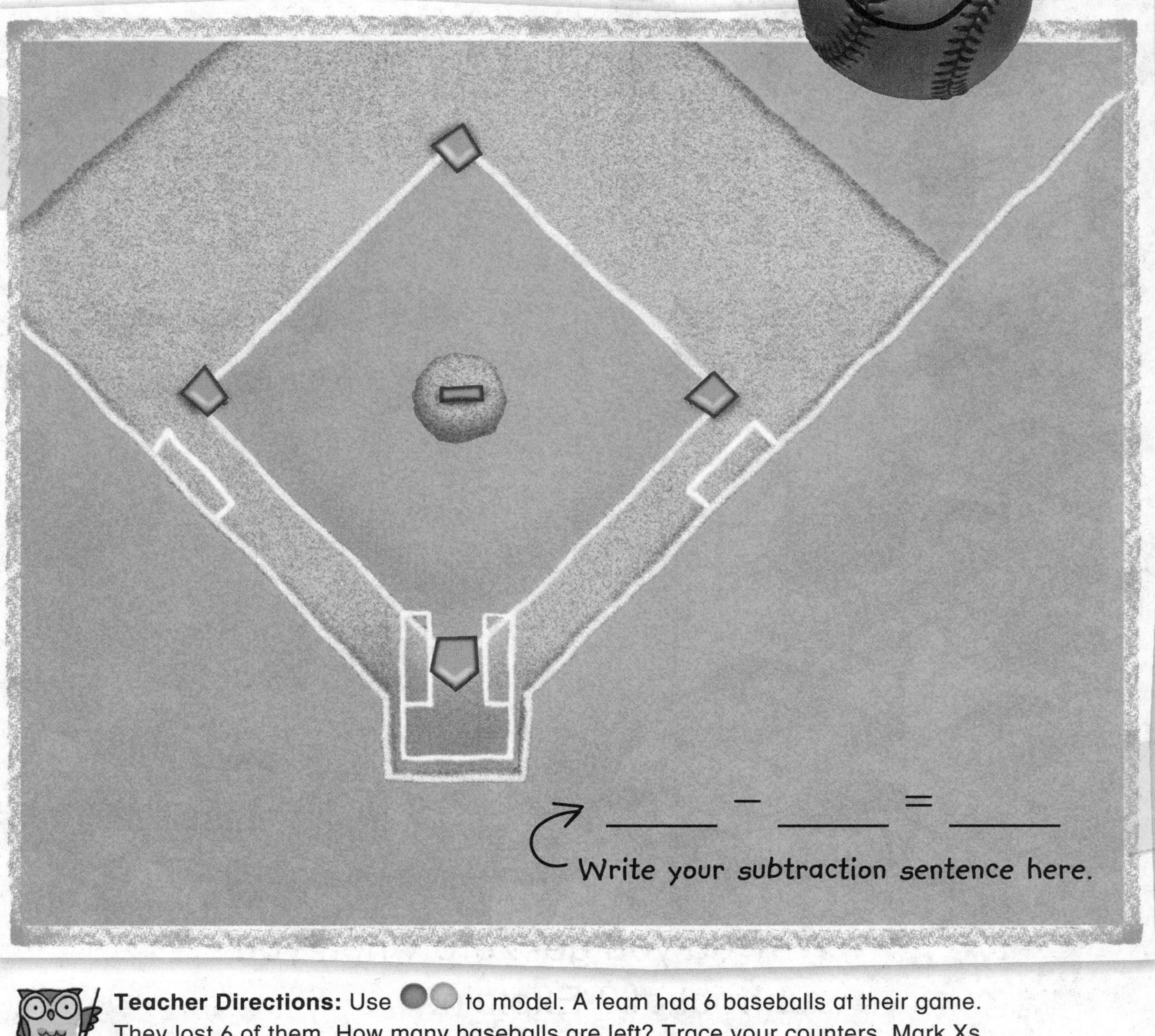

_____ − _____ = _____

Write your subtraction sentence here.

Teacher Directions: Use counters to model. A team had 6 baseballs at their game. They lost 6 of them. How many baseballs are left? Trace your counters. Mark Xs on the baseballs that are lost. Write the subtraction number sentence.

See and Show

Processes & Practices

When you subtract 0, you have the same number left.

4 − 0 = 4

When you subtract all, you have 0 left.

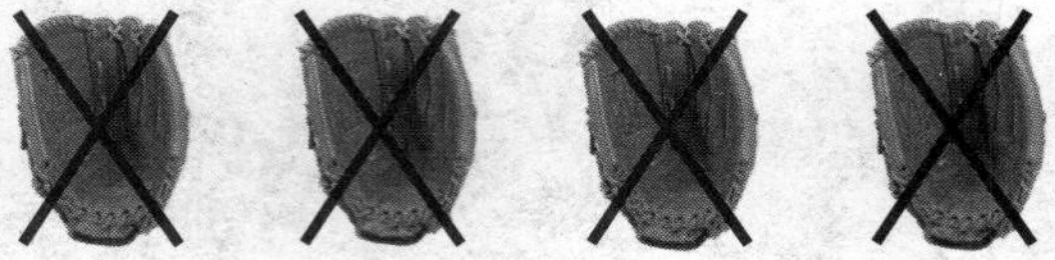

4 − 4 = 0

Subtract.

1. 

5 − 5 = ______

2. 

8 − 0 = ______

3. 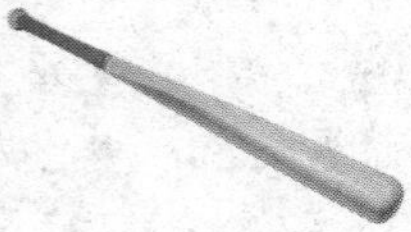

1 − 0 = ______

4. 

3 − 3 = ______

Talk Math Why do you get zero when you subtract all? Explain.

Name ....

On My Own

Subtract.

5. 

3 − 0 = ______

6. 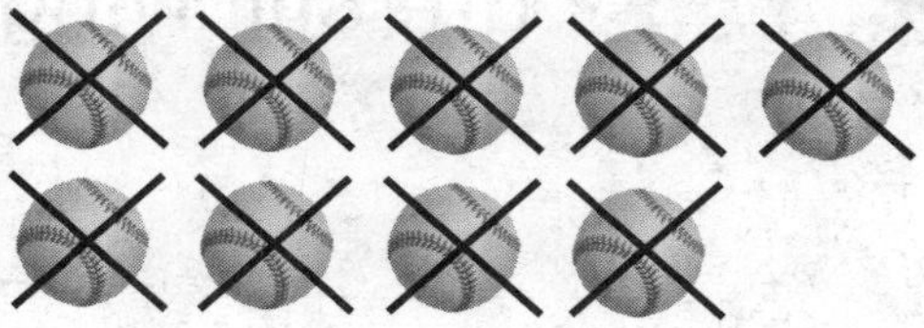

9 − 9 = ______

7. 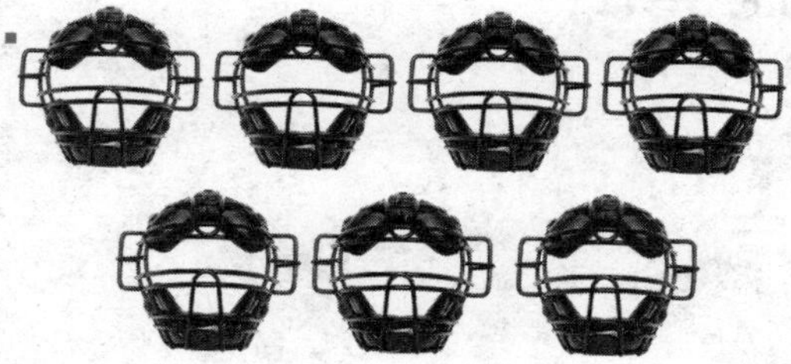

7 − 0 = ______

8. 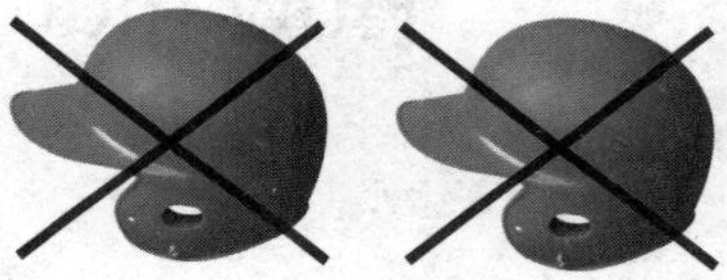

2 − 2 = ______

9. 8 − 0 = ______

10. 3 − 3 = ______

11. 6 − 6 = ______

12. 4 − 0 = ______

13. 7 − 7 = ______

14. 1 − 1 = ______

15. 9 − 0 = ______

16. 5 − 0 = ______

Real World!

Problem Solving

Write a subtraction number sentence.

17. There are 9 bears lying down. 0 of the bears leave. How many bears are still lying down?

_____ – _____ = _____ bears

18. A koala bear finds 7 leaves on the ground. The koala eats 7 of the leaves. How many leaves are left?

_____ – _____ = _____ leaves

HOT Problem A parrot has 6 baby birds in a nest. 0 baby birds fly away. The answer is 6 baby birds. What is the question?

Name ....

My Homework

Lesson 4
Subtract 0 and All

Homework Helper

eHelp Need help? connectED.mcgraw-hill.com

When you subtract 0, you have the same number left.

4 − 0 = 4

When you subtract all, you have 0 left.

4 − 4 = 0

Practice

Subtract.

1.

5 − 0 = ______

2.

6 − 6 = ______

3.

3 − 3 = ______

4.

7 − 0 = ______

Subtract.

5. 1 − 0 = ______

6. 9 − 9 = ______

7. 7 − 7 = ______

8. 6 − 0 = ______

9. 5 − 0 = ______

10. 8 − 8 = ______

11. There are 8 parrots on a branch. 0 parrots fly away. How many parrots are left on the branch?

______ parrots

Test Practice

12. There are 9 lizards on a leaf. 9 of them crawl away. How many lizards are still on the leaf?

18	9	5	0
○	○	○	○

Math at Home Give your child 3 objects. Have them use the objects to show 3 − 0 and 3 − 3.

Name ....

Vertical Subtraction

Lesson 5

ESSENTIAL QUESTION
How do you subtract numbers?

Explore and Explain

Watch Tools Vocab

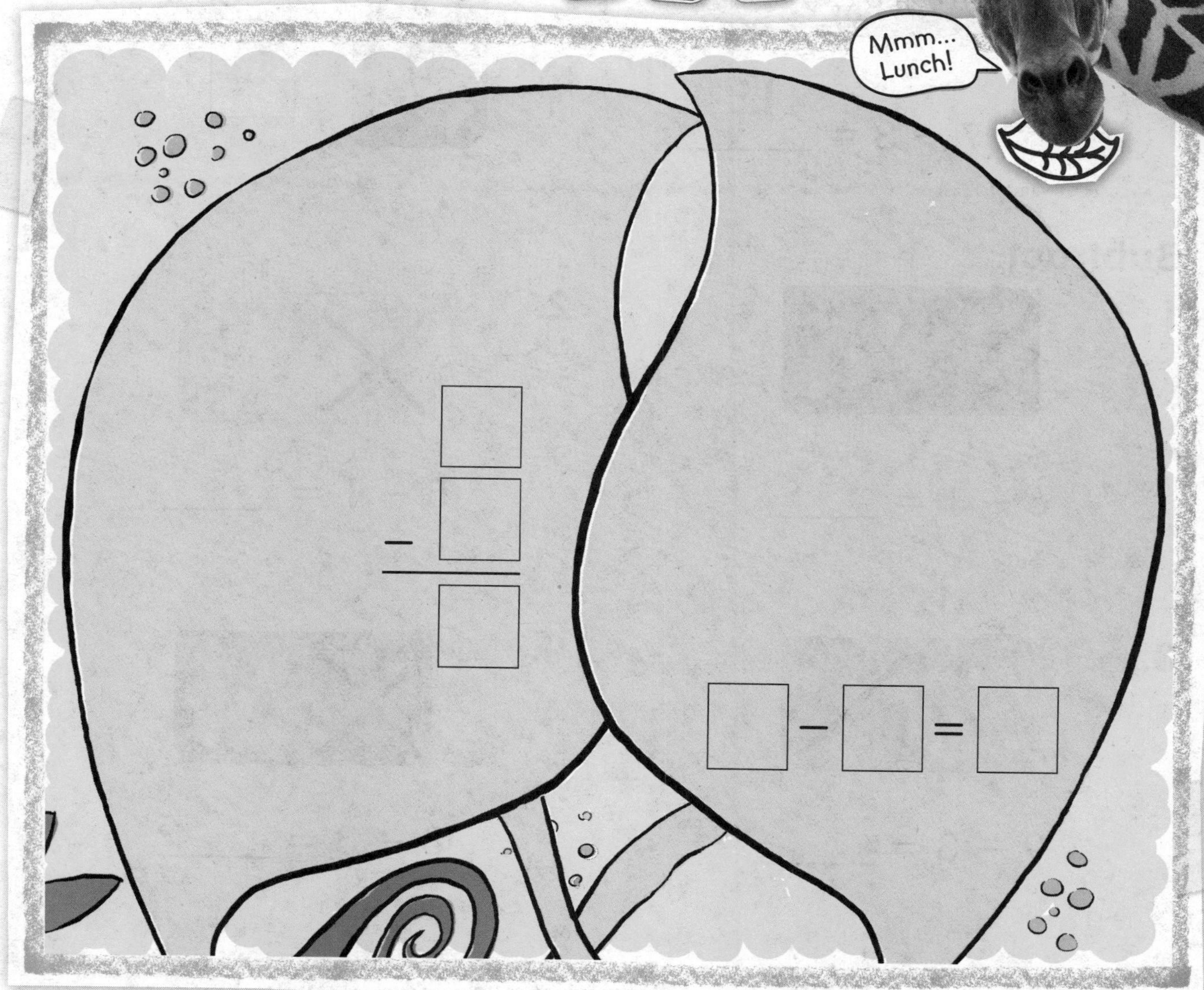

Teacher Directions: Use counters to model. 4 bugs landed on each leaf. Then 3 bugs flew away from each of the leaves. How many bugs are left on each leaf? Trace the counters you used. Draw Xs on the counters to show the bugs that flew away. Write the subtraction number sentences.

See and Show

Processes & Practices

You can subtract across or you can subtract down. When the same numbers are used, the answer is the same.

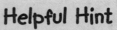

Helpful Hint
You can write subtraction number sentences two ways.

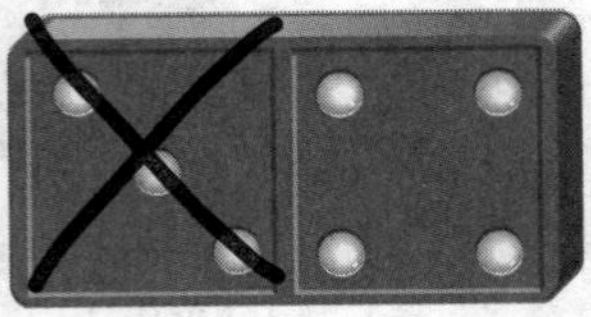

7 − 3 = 4

$$\begin{array}{r} 7 \\ -\ 3 \\ \hline \boxed{4} \end{array}$$

Subtract.

1\.

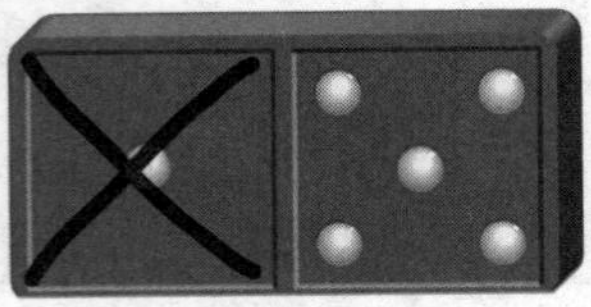

6 − 1 = ______

2\.

5 − 1 = ______

3\.

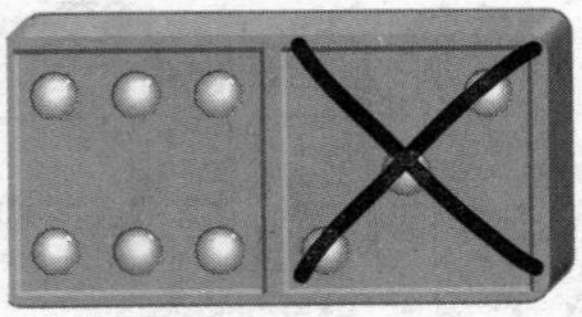

9 − 3 = ______

4\.

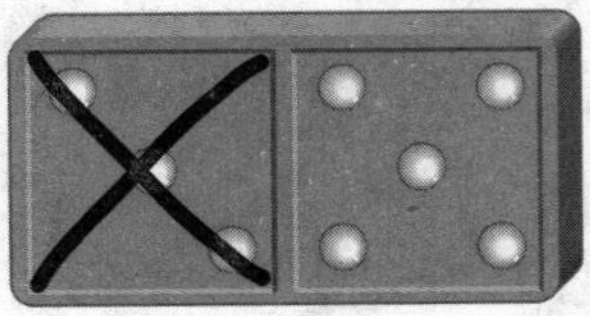

8 − 3 = ______

Talk Math How is subtracting down like subtracting across?

Name ....

On My Own

Subtract.

5.

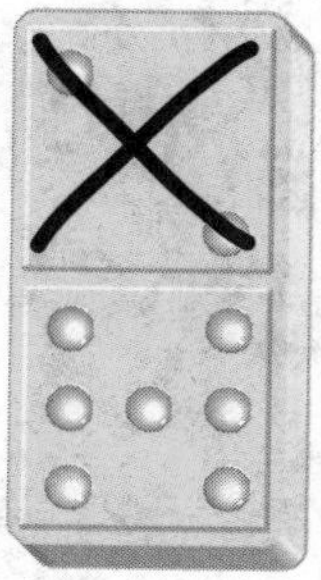

$$\begin{array}{r} 9 \\ -\ 2 \\ \hline \square \end{array}$$

6.

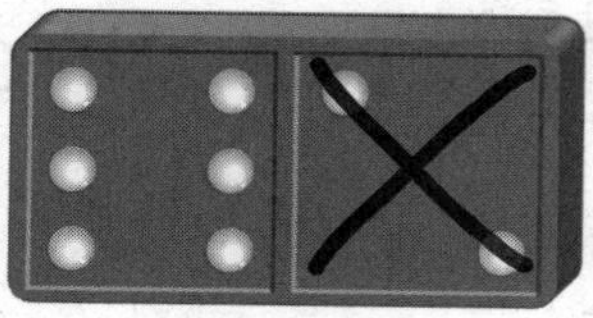

8 − 2 = ______

7.

$$\begin{array}{r} 4 \\ -\ 2 \\ \hline \square \end{array}$$

8.

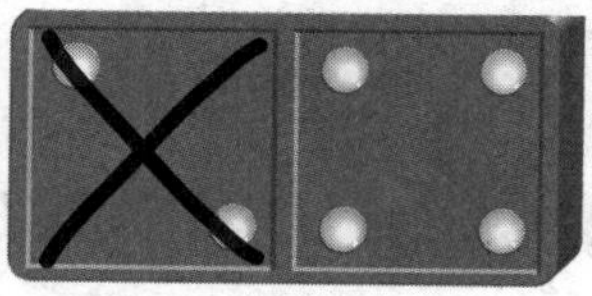

6 − 2 = ______

9.

7 − 5 = ______

10.

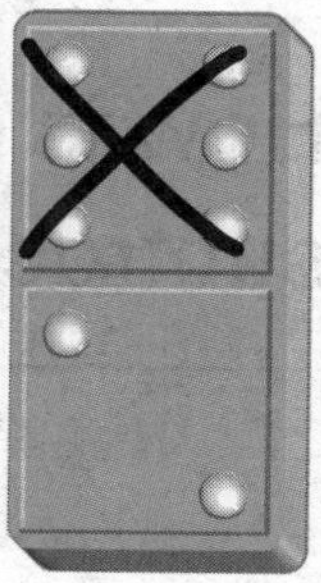

$$\begin{array}{r} 8 \\ -\ 6 \\ \hline \square \end{array}$$

11. 5 − 4 = ______

12. 8 − 5 = ______

13.

$$\begin{array}{r} 9 \\ -\ 1 \\ \hline \square \end{array}$$

14.

$$\begin{array}{r} 6 \\ -\ 6 \\ \hline \square \end{array}$$

Problem Solving

Processes & Practices

Write a subtraction number sentence.

15. There are 8 zebras eating grass. 2 zebras stop eating. How many zebras are still eating grass?

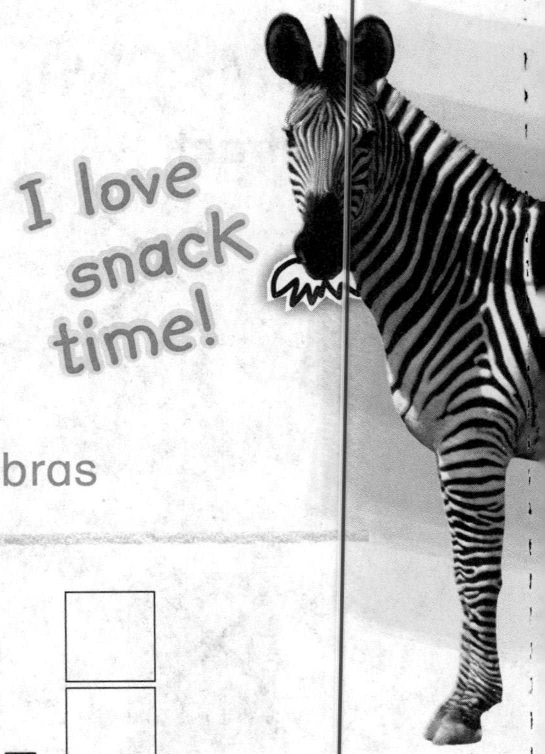

_____ − _____ = _____ zebras

16. There are 9 leopards in a field. 2 leopards leave. How many leopards are still in the field?

$$\begin{array}{r} \square \\ -\ \square \\ \hline \square \end{array}$$ leopards

Write Math How is subtracting down different from subtracting across?

____

____

____

Name ....

My Homework

Lesson 5
Vertical Subtraction

Homework Helper

eHelp Need help? connectED.mcgraw-hill.com

You can subtract across, or you can subtract down.

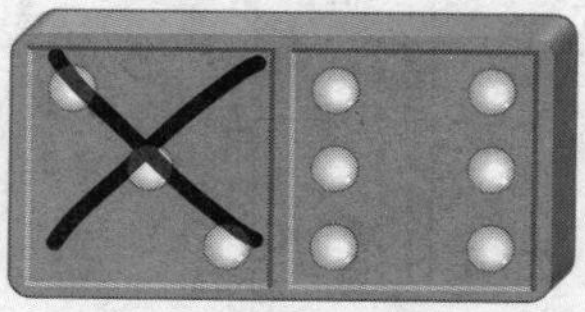

$9 - 3 = 6$

$$\begin{array}{r} 9 \\ -\ 3 \\ \hline 6 \end{array}$$

Practice

Subtract.

1.

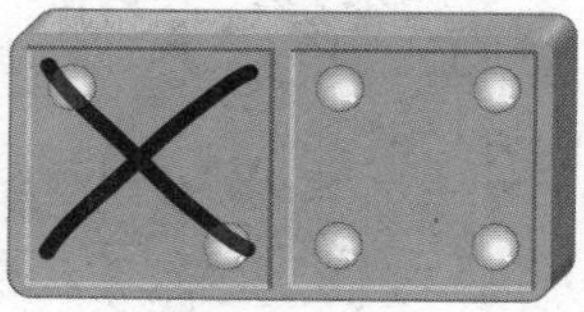

$6 - 2 =$ ______

2.

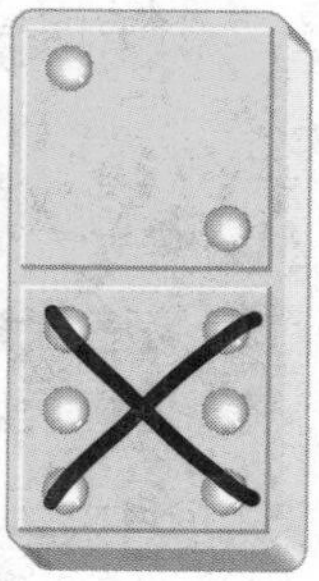

$$\begin{array}{r} 8 \\ -\ 6 \\ \hline \square \end{array}$$

3.

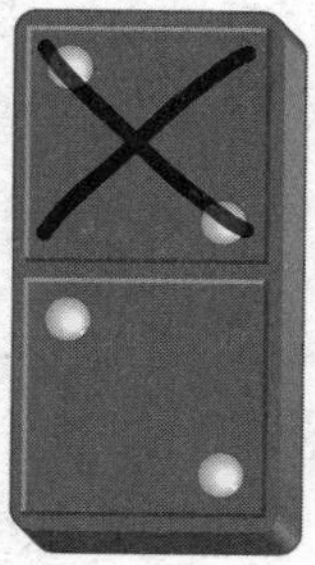

$$\begin{array}{r} 4 \\ -\ 2 \\ \hline \square \end{array}$$

4.

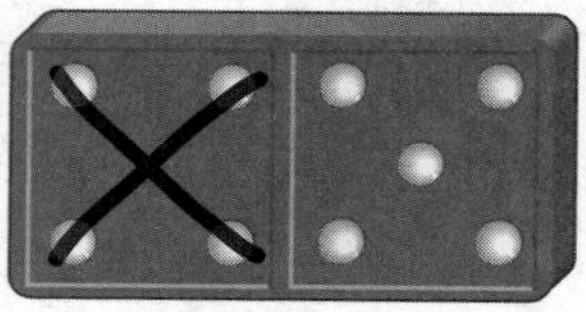

$9 - 4 =$ ______

Subtract.

5. 3 − 1 = ______

6. 9 − 6 = ______

7.
$$\begin{array}{r} 7 \\ -\ 2 \\ \hline \square \end{array}$$

8.
$$\begin{array}{r} 8 \\ -\ 1 \\ \hline \square \end{array}$$

9.
$$\begin{array}{r} 2 \\ -\ 2 \\ \hline \square \end{array}$$

10. There are 7 mangos growing on a tree. Diego picks 3 of the mangos. How many mangos are still on the tree?

$$\begin{array}{r} \square \\ -\ \square \\ \hline \square \end{array}$$ mangos

Test Practice

11. Which number sentence is shown?

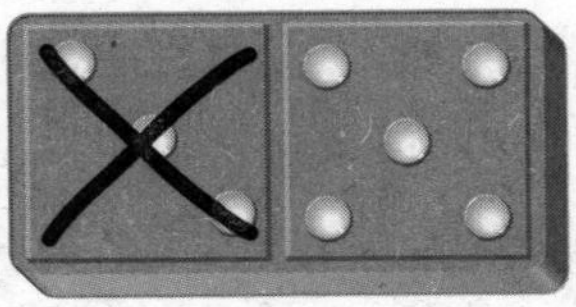

○ 5 − 3 = 2

○ 8 − 3 = 5

○ 3 − 3 = 0

○ 3 + 5 = 8

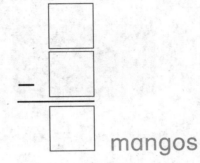

Math at Home Use 9 small objects. Show subtraction by taking some objects away. Have your child write the subtraction number sentence vertically and horizontally.

Name ....

Check My Progress

Vocabulary Check

Draw lines to match.

1. subtract	–
2. minus sign	To take away a part from the whole.
3. subtraction number sentence	The answer in a subtraction problem.
4. difference	6 – 5 = 1

Concept Check

Tell a number story. Write how many are left.

5.

How many hippos are left in the water?

_____ hippos

Write a subtraction number sentence.

6. 

____ ○ ____ ○ ____

Subtract.

7. 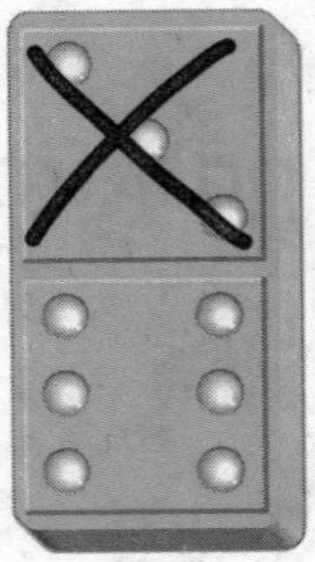

$$\begin{array}{r} 9 \\ -\ 3 \\ \hline \square \end{array}$$

8. 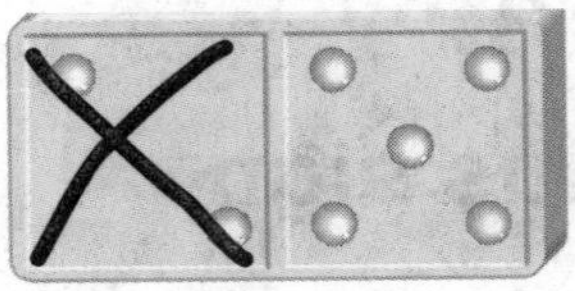

7 − 2 = ____

9. There are 5 monkeys on a tree.
0 monkeys get out of the tree. How many monkeys are left on the tree?

____ monkeys

Test Practice

10. There are 9 owls. 3 owls fly away. How many owls are left?

6 owls ○ 11 owls ○ 7 owls ○ 12 owls ○

Name ________________________________

Problem Solving

STRATEGY: Draw a Diagram

Lesson 6

ESSENTIAL QUESTION
How do you subtract numbers?

Abby has 5 erasers. She gives Matt 2 erasers. How many erasers does Abby have left?

1 Understand

Underline what you know.
Circle what you need to find.

2 Plan

How will I solve the problem?

3 Solve

I will draw a diagram.

3 erasers

4 Check

Is my answer reasonable? Explain.

Practice the Strategy

Lila has 8 toys. She lets Rex play with 3 of the toys. How many toys does Lila have left?

1 Understand Underline what you know. Circle what you need to find.

2 Plan How will I solve the problem?

3 Solve I will...

_____ toys

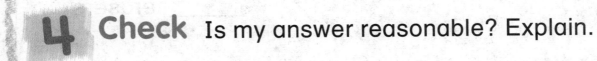

4 Check Is my answer reasonable? Explain.

Name ....

Processes & Practices

Apply the Strategy

Draw a diagram to solve.

1. Nick has 6 cherries.
   He eats 3 of the cherries.
   How many cherries are left?

   ______ cherries

2. Alberto buys 7 apples.
   He eats 1 apple. How many apples does he have left?

   ______ apples

3. There are 7 oranges. Miles eats some of the oranges.
   There are 5 oranges left. How many oranges did Miles eat?

   ______ oranges

Review the Strategies

Choose a strategy

- Draw a diagram.
- Act it out.
- Write a number sentence.

4. Lena has 6 books. She gives 2 books to her sister. How many books does Lena have left?

_____ books

5. Jessica catches 4 frogs. 3 of the frogs hop away. How many frogs does Jessica have now?

_____ frog

6. Marcos ate 5 crackers. Shani ate some of the crackers. Together they ate 9 crackers. How many crackers did Shani eat?

_____ crackers

Name ______________________________

My Homework

Lesson 6

Problem Solving: Draw a Diagram

Homework Helper

eHelp **Need help?** connectED.mcgraw-hill.com

Asia is watching 6 birds sitting on a tree. 3 of the birds fly away. How many birds are still on the tree?

1 Understand Underline what you know. Circle what you need to find.

2 Plan How will I solve the problem?

3 Solve I will draw a diagram.

3 birds

4 Check Is my answer reasonable?

Real World Problem Solving

Underline what you know. Circle what you need to find. Draw a diagram to solve.

1. There are 9 frogs on a tree. 4 of the frogs hop away. How many frogs are left on the tree?

_____ frogs

2. Max sees 7 butterflies on a flower. 5 of them fly away. How many butterflies are left?

I'm all a-flutter over math!

_____ butterflies

3. There are 8 pandas sitting near a tree. 4 of them leave. How many pandas are still near the tree?

_____ pandas

Math at Home Give your child a simple subtraction problem and have him or her solve it by drawing a picture.

Name

Compare Groups

Lesson 7

ESSENTIAL QUESTION
How do you subtract numbers?

Explore and Explain

Watch Tools

Teacher Directions: Look at the picture. Place [cube] on each bird. Place [cube] on each monkey. Count and write how many of each animal. Circle the number that shows more.

See and Show

You can subtract to **compare** groups.

There are 6 zebras. There are 2 elephants. How many more zebras are there than elephants?

Zebras rule and elephants drool!

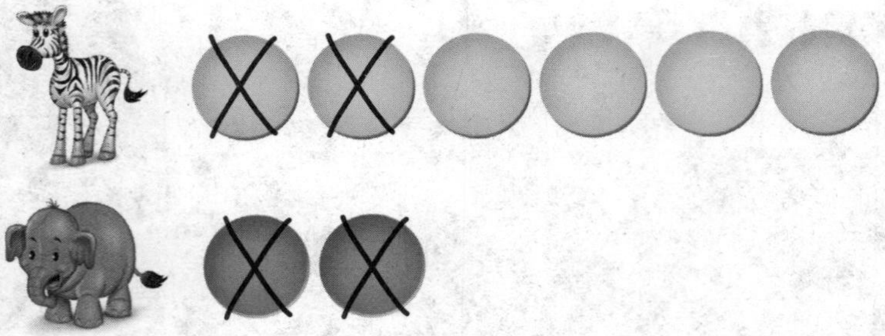

6 – 2 = 4 more zebras

4 fewer elephants

Use ●●. Write a subtraction number sentence. Write how many more or fewer.

1. There are 7 giraffes. There are 2 rhinos. How many more giraffes are there than rhinos?

   ____ – ____ = ____ more giraffes

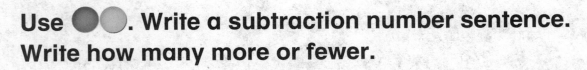

Talk Math What happens when you compare equal groups?

Name ....

On My Own

Can you help?

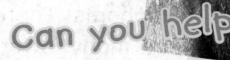

Use ●●. Write a subtraction number sentence. Write how many more or fewer.

2. There are 5 leopards. There are 3 gorillas. How many fewer gorillas are there than leopards?

_____ − _____ = _____ fewer gorillas

3. 9 butterflies are black. 5 butterflies are yellow. How many fewer butterflies are yellow than black?

_____ − _____ = _____ fewer butterflies

4. There are 6 tigers. There are 2 cheetahs. How many more tigers are there than cheetahs?

_____ − _____ = _____ more tigers

Problem Solving

Processes & Practices

5. There are 8 elephants. There are 4 lions. How many fewer lions are there than elephants?

______ fewer lions

6. There are 9 monkeys and 8 parrots in a tree. How many more monkeys are there than parrots?

______ more monkey

Write Math How can you use subtraction to compare groups?

Name ....

My Homework

Lesson 7
Compare Groups

Homework Helper

eHelp Need help? connectED.mcgraw-hill.com

There are 4 bears. There are 3 foxes. How many more bears are there than foxes?

4 − 3 = 1 more bear

Practice

Write a subtraction number sentence.
Write how many more or fewer.

1. There are 4 snakes. There are 2 chimps. How many fewer chimps are there than snakes?

   ______ − ______ = ______ fewer chimps

2. Stella ate 3 bananas. Seth ate 2 bananas. How many fewer bananas did Seth eat than Stella?

   ______ − ______ = ______ fewer banana

Write a subtraction number sentence.
Write how many more or fewer.

3. There are 5 frogs in a pond and 3 frogs on the land. How many more frogs are in a pond than on the land?

_____ − _____ = _____ more frogs

4. Jesse went fishing. He caught 7 fish in the morning and 6 fish in the afternoon. How many fewer fish did Jesse catch in the afternoon than in the morning?

_____ − _____ = _____ fewer fish

Vocabulary Check

Vocab

Circle the correct answer.

compare **add**

5. Subtract two different groups so you can _____________ them to find which group has more and which has fewer.

Math at Home Take a walk outside. Collect leaves or other objects found in nature. Create two groups with the items showing less than 9 items in each group. Have your child tell how many more or how many fewer.

Name ______________________________

Subtract from 4 and 5

Lesson 8

ESSENTIAL QUESTION
How do you subtract numbers?

Where's my snorkel?

Explore and Explain

Watch Tools

_____ – _____ = _____
Write your subtraction sentence here.

Teacher Directions: Use cubes to model. There are 4 crocodiles in a lake. 3 crocodiles get out. How many crocodiles are still in the lake? Trace your cubes and mark Xs on the cubes to show the crocodiles that get out. Write the subtraction number sentence.

See and Show

Processes & Practices

You can subtract from 4 and 5.

Subtract 2 from 4.

4 − 2 = 2 The difference is 2.

Subtract 1 from 5.

5 − 1 = 4 The difference is 4.

Start with 4 cubes. Subtract some cubes. Cross out cubes. Write different ways to subtract from 4.

Subtract from 4

1.  4 − ____ = ____

2.  4 − ____ = ____

3.  4 − ____ = ____

4.  4 − ____ = ____

Talk Math What does difference mean in subtraction?

Name ....

On My Own

Start with 5 cubes. Subtract some cubes. Cross out cubes. Write different ways to subtract from 5.

Subtract from 5

5.  5 − ____ = ____

6.  5 − ____ = ____

7.  5 − ____ = ____

8. 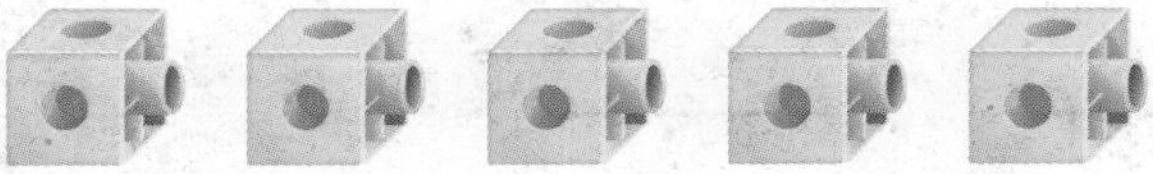 5 − ____ = ____

9.  5 − ____ = ____

Subtract. Use Work Mat 3 and cubes.

10. 4 − 1 = ____

11. 5 − 2 = ____

12. 5 − 5 = ____

13. 4 − 2 = ____

14.
  5
− 3
____

15.
  5
− 0
____

16.
  4
− 4
____

Problem Solving

Processes & Practices

Write a subtraction number sentence.

17. Yuan draws 4 hippos. He crosses out 2.
How many hippos are there now?

_____ − _____ = _____ hippos

18. Billy draws 5 lions. He crosses out 1.
How many lions are there now?

_____ − _____ = _____ lions

HOT Problem Isabel wrote this subtraction sentence. Tell why Isabel is wrong. Make it right.

5 - 2 = 4

Name ....

My Homework

Lesson 8

Subtract from 4 and 5

Homework Helper

eHelp

Need help? connectED.mcgraw-hill.com

You can subtract from 4 and 5.

$4 - 1 = 3$

$5 - 3 = 2$

Practice

Write different ways to subtract from 4 and 5.

1. 4 – ______ = ______
2. 4 – ______ = ______
3. 4 – ______ = ______
4. 4 – ______ = ______
5. 5 – ______ = ______
6. 5 – ______ = ______
7. 5 – ______ = ______
8. 5 – ______ = ______

Subtract.

9. 5 – 3 = ______
10. 4 – 4 = ______

Subtract.

11. 4 − 2 = ______

12. 5 − 1 = ______

13. 4 − 1 = ______

14. 5 − 4 = ______

15. 5 − 0 = ______

16. 4 − 3 = ______

17. $\begin{array}{r} 5 \\ -\ 5 \\ \hline \end{array}$

18. $\begin{array}{r} 5 \\ -\ 2 \\ \hline \end{array}$

19. $\begin{array}{r} 4 \\ -\ 0 \\ \hline \end{array}$

20. Chad rents 5 movies. He watches 3 of the movies. How many movies does he have left to watch?

______ movies

Test Practice

21. 4 − 4 = ______

0	1	6	8
○	○	○	○

Math at Home Give your child 5 objects. Have him or her subtract different numbers from 4 or 5 and tell the difference.

Name

Subtract from 6 and 7

Lesson 9

ESSENTIAL QUESTION

How do you subtract numbers?

Explore and Explain

Watch Tools

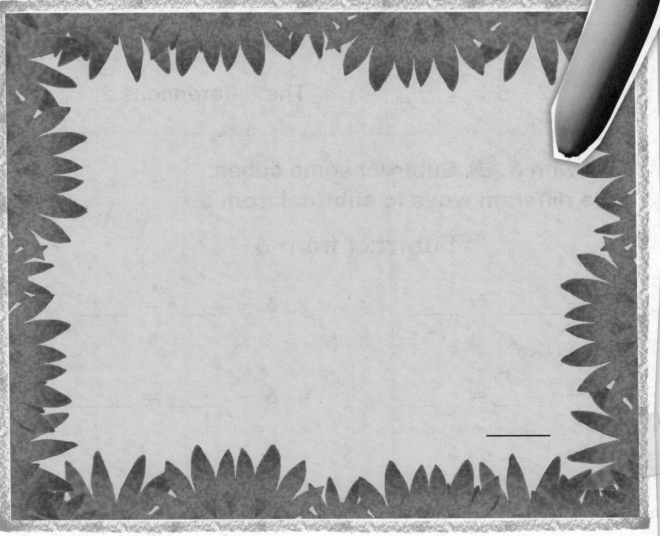

Teacher Directions: Use [cube] to model. There are 7 apes eating bananas. 3 apes stop eating bananas. How many apes are still eating bananas? Trace the cubes you used. Mark Xs on the cubes to show the apes that stop eating bananas. Write the number.

See and Show

Processes & Practices

You can subtract from 6 and 7.

Subtract 3 from 6.

6 – 3 = 3 The difference is 3.

Subtract 5 from 7.

7 – 5 = 2 The difference is 2.

Start with 6 . Subtract some cubes.
Write different ways to subtract from 6.

Subtract from 6

1. 6 – _____ = _____	2. 6 – _____ = _____
3. 6 – _____ = _____	4. 6 – _____ = _____
5. 6 – _____ = _____	6. 6 – _____ = _____

Talk Math How could you use  to show subtraction?

Name ______________________

On My Own

Start with 7 cubes. Subtract some cubes.
Write different ways to subtract from 7.

Subtract from 7

7. 7 − ____ = ____
8. 7 − ____ = ____
9. 7 − ____ = ____
10. 7 − ____ = ____
11. 7 − ____ = ____
12. 7 − ____ = ____

Subtract. Use Work Mat 3 and cubes.

13. 7 − 5 = ____
14. 6 − ____ = 4
15. 7 − ____ = 6
16. 6 − ____ = 0

17. $\begin{array}{r} 7 \\ -\ 0 \\ \hline \end{array}$

18. $\begin{array}{r} 7 \\ -\ 6 \\ \hline \end{array}$

19. $\begin{array}{r} 6 \\ -\ 5 \\ \hline \end{array}$

Problem Solving

Processes &Practices

Write a subtraction number sentence to solve.

20. 7 rhinos get a drink from a pond. 3 of them stop drinking. How many rhinos are still drinking?

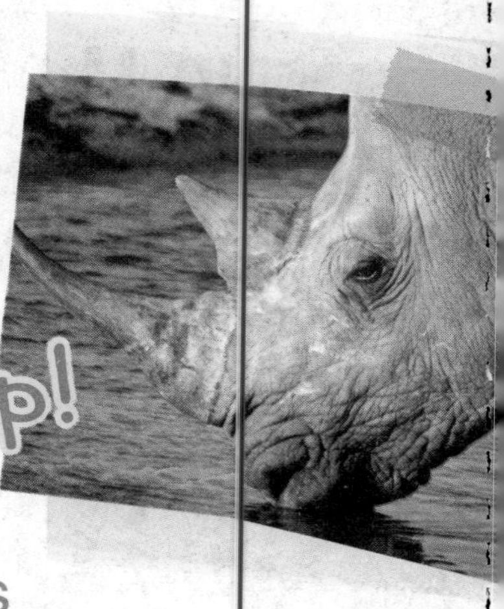

_____ – _____ = _____ rhinos

21. 7 deer are walking in a field. 5 deer lie down. How many deer are still walking?

_____ – _____ = _____ deer

Write Math What happens to the number of objects in a group when they are subtracted? Explain.

Name ______________________________

My Homework

Lesson 9

Subtract from 6 and 7

Homework Helper

eHelp

Need help? connectED.mcgraw-hill.com

You can subtract from 6 and 7.

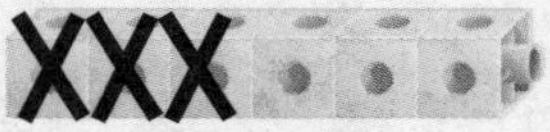

6 − 3 = 3

7 − 2 = 5

Practice

Write different ways to subtract from 6 and 7.

1. 6 − ______ = ______

2. 6 − ______ = ______

3. 6 − ______ = ______

4. 6 − ______ = ______

5. 7 − ______ = ______

6. 7 − ______ = ______

7. 7 − ______ = ______

8. 7 − ______ = ______

Subtract.

9. 6 − 1 = ______

10. 7 − 4 = ______

Subtract.

11. 7 − 1 = ______

12. 7 − 4 = ______

13. 6 − 5 = ______

14. 7 − 7 = ______

15. $\begin{array}{r} 7 \\ -\ 6 \\ \hline \end{array}$

16. $\begin{array}{r} 7 \\ -\ 0 \\ \hline \end{array}$

17. $\begin{array}{r} 6 \\ -\ 1 \\ \hline \end{array}$

18. There are 7 ants on a log. 5 of the ants crawl away. How many ants are left on the log?

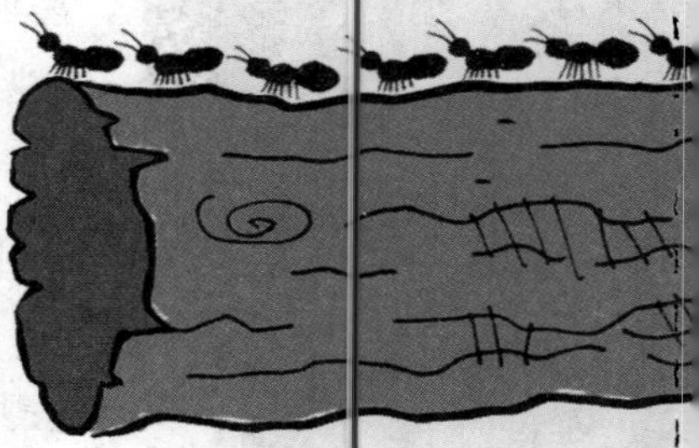

______ ants

Test Practice

19. Ben sees 6 parrots on a branch. 2 of them fly away. How many parrots are left on the branch?

0 parrots ○	2 parrots ○	3 parrots ○	4 parrots ○

Math at Home Collect a group of 7 objects. Have your child show you how to subtract from 7. Have him or her write a subtraction number sentence.

Name ....

Check My Progress

Vocabulary Check

Circle the correct answer.

difference **compare**

1. The words more and fewer can be used to ____________ the number of objects in two different groups.

Concept Check

Write a subtraction number sentence.

2. 

_____ − _____ = _____

3. 

_____ − _____ = _____

Subtract.

4. $\begin{array}{r} 6 \\ -\ 4 \\ \hline \end{array}$

5. $\begin{array}{r} 4 \\ -\ 4 \\ \hline \end{array}$

6. $\begin{array}{r} 5 \\ -\ 1 \\ \hline \end{array}$

Subtract.

7. 5 − 4 = ____

8. 6 − 0 = ____

9. 6 − 2 = ____

10. 7 − 7 = ____

11. 7 − 4 = ____

12. 4 − 2 = ____

13. There are 6 beetles. There are 4 butterflies. How many fewer butterflies are there than beetles?

____ − ____ = ____ fewer butterflies

Test Practice

14. Rachel sees 7 caterpillars on a leaf. 5 of them crawl away. How many caterpillars are on the leaf now?

- ○ 12 caterpillars
- ○ 11 caterpillars
- ○ 3 caterpillars
- ○ 2 caterpillars

Name ....

Subtract from 8

Lesson 10

ESSENTIAL QUESTION
How do you subtract numbers?

Explore and Explain

Watch Tools

____ − ____ = ____
Write your subtraction sentence here.

Teacher Directions: Use [cube] to model. 8 koalas are on a tree. 6 of the koalas climb away. How many koalas are still on the tree? Trace your cubes and mark Xs on the number of koalas that climb away. Write the subtraction number sentence.

See and Show

Processes &Practices

There are many ways to subtract from 8.

Subtract 4 from 8.

8 − 4 = 4   The difference is 4.

Subtract 6 from 8.

8 − 6 = 2   The difference is 2.

Start with 8 . Subtract some cubes. Write different ways to subtract from 8.

Subtract from 8

1. 8 − _____ = _____	2. 8 − _____ = _____
3. 8 − _____ = _____	4. 8 − _____ = _____
5. 8 − _____ = _____	6. 8 − _____ = _____

Talk Math How do you know 8 − 5 = 3? Explain.

Name ____________________

I go bananas over subtraction!

On My Own

Use Work Mat 3 and [cube]. Subtract.

7. 7 − 3 = ______

8. 8 − 5 = ______

9. 8 − 1 = ______

10. 8 − 7 = ______

11. 8 − 0 = ______

12. 8 − 6 = ______

13. 8 − 2 = ______

14. 8 − 4 = ______

15. 8 − 8 = ______

16. 5 − 4 = ______

17. $\begin{array}{r} 8 \\ -\ 6 \\ \hline \end{array}$

18. $\begin{array}{r} 6 \\ -\ 2 \\ \hline \end{array}$

19. $\begin{array}{r} 7 \\ -\ 6 \\ \hline \end{array}$

20. $\begin{array}{r} 8 \\ -\ 0 \\ \hline \end{array}$

21. $\begin{array}{r} 6 \\ -\ 5 \\ \hline \end{array}$

22. $\begin{array}{r} 8 \\ -\ 3 \\ \hline \end{array}$

Problem Solving

Real World!

Processes & Practices

Write a subtraction number sentence to solve.

23. There are 8 giraffes drinking water. 3 giraffes go away. How many giraffes are left drinking water?

_____ − _____ = _____ giraffes

24. There are 6 wolves playing in a field. 2 wolves run away. How many wolves are left in the field?

_____ − _____ = _____ wolves

HOT Problem Nia wrote this subtraction number sentence. Tell why Nia is wrong. Make it right.

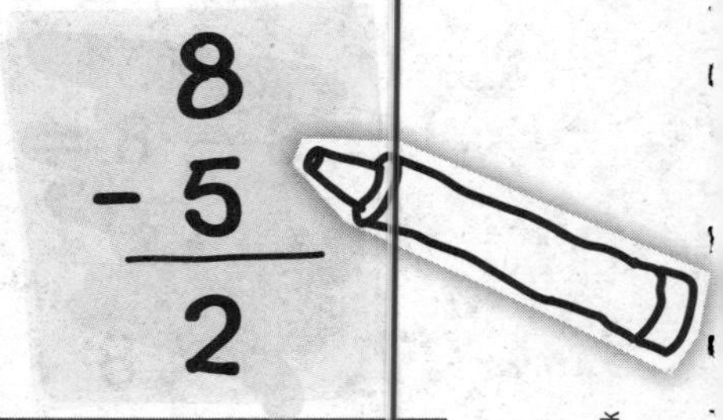

Name ______________________________

My Homework

Lesson 10
Subtract from 8

Homework Helper

eHelp **Need help?** connectED.mcgraw-hill.com

There are many ways to subtract from 8.

8 − 7 = 1

8 − 4 = 4

Practice

Write different ways to subtract from 8.

1. 8 − ______ = ______

2. 8 − ______ = ______

3. 8 − ______ = ______

4. 8 − ______ = ______

5. 8 − ______ = ______

6. 8 − ______ = ______

7. 8 − ______ = ______

8. 8 − ______ = ______

Subtract.

9. 8 − 3 = ______

10. 8 − 4 = ______

Subtract.

11. $8 - 2 =$ ______

12. $7 - 4 =$ ______

13. $8 - 7 =$ ______

14. $5 - 5 =$ ______

15. $\begin{array}{r} 6 \\ -\ 3 \\ \hline \end{array}$

16. $\begin{array}{r} 7 \\ -\ 5 \\ \hline \end{array}$

17. $\begin{array}{r} 8 \\ -\ 8 \\ \hline \end{array}$

18. FaShaun has 8 zebra stickers. He gives his friend 3 of them. How many stickers does FaShaun have left?

______ stickers

Test Practice

19. There are 8 rhinos getting a drink. 6 of the rhinos walk away. How many rhinos are still getting a drink?

0 rhinos	2 rhinos	6 rhinos	14 rhinos
○	○	○	○

Math at Home Collect a group of 8 objects. Have your child show you all of the numbers you can subtract from 8. Have him or her write each subtraction number sentence.

Name

Subtract from 9

Lesson 11

ESSENTIAL QUESTION
How do you subtract numbers?

Explore and Explain

Watch Tools

Catch me if you can!
___ − ___ = ___
Write your subtraction sentence here.

Teacher Directions: Use [cube] to model. There are 9 bugs on a flower. 2 of the bugs fly away. How many bugs are left? Trace your cubes and mark Xs on the bugs that fly away. Write the subtraction number sentence.

See and Show

There are many ways to subtract from 9.

Subtract 3 from 9.

9 − 3 = 6 The difference is 6.

Subtract 5 from 9.

9 − 5 = 4 The difference is 4.

Start with 9 **. Subtract some cubes.**
Write different ways to subtract from 9.

Subtract from 9

1. 9 − ____ = ____
2. 9 − ____ = ____
3. 9 − ____ = ____
4. 9 − ____ = ____
5. 9 − ____ = ____
6. 9 − ____ = ____
7. 9 − ____ = ____
8. 9 − ____ = ____

Talk Math How do you know when to subtract?

Name ____________

Let's take a closer look!

On My Own

Use Work Mat 3 and [cube]. Subtract.

9. 7 − 5 = ______

10. 9 − 3 = ______

11. 9 − 2 = ______

12. 5 − 3 = ______

13. 9 − 0 = ______

14. 9 − 5 = ______

15. 6 − 2 = ______

16. 9 − 4 = ______

17. 9 − 8 = ______

18. 7 − 7 = ______

19. $\begin{array}{r} 7 \\ -\ 0 \\ \hline \end{array}$

20. $\begin{array}{r} 9 \\ -\ 6 \\ \hline \end{array}$

21. $\begin{array}{r} 9 \\ -\ 9 \\ \hline \end{array}$

22. $\begin{array}{r} 9 \\ -\ 7 \\ \hline \end{array}$

23. $\begin{array}{r} 9 \\ -\ 1 \\ \hline \end{array}$

24. $\begin{array}{r} 9 \\ -\ 2 \\ \hline \end{array}$

Real World

Problem Solving

Processes & Practices

Solve. Use Work Mat 3 and [cube] if needed.

25. Andre had 9 hats. He lost 5 of them. How many hats does Andre have left?

_____ hats

26. 7 ladybugs are climbing on a fence. 3 ladybugs fly away. How many ladybugs are left?

_____ ladybugs

Write Math Why is the answer to a subtraction problem called the difference?

____

____

____

Name ______________________________

My Homework

Lesson 11
Subtract from 9

Homework Helper

eHelp Need help? connectED.mcgraw-hill.com

There are many ways to subtract from 9.

$9 - 7 = 2$

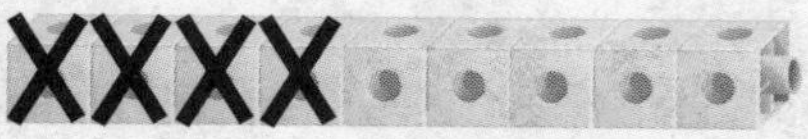

$9 - 4 = 5$

Practice

Write different ways to subtract from 9.

1. 9 − ______ = ______
2. 9 − ______ = ______
3. 9 − ______ = ______
4. 9 − ______ = ______
5. 9 − ______ = ______
6. 9 − ______ = ______
7. 9 − ______ = ______
8. 9 − ______ = ______

Subtract.

9. 9 − 3 = ______
10. 9 − 4 = ______

Subtract.

11. $9 - 2 =$ ______

12. $9 - 6 =$ ______

13. $7 - 3 =$ ______

14. $9 - 5 =$ ______

15. $\begin{array}{r} 9 \\ -\ 9 \\ \hline \end{array}$

16. $\begin{array}{r} 5 \\ -\ 2 \\ \hline \end{array}$

17. $\begin{array}{r} 9 \\ -\ 0 \\ \hline \end{array}$

18. There are 9 frogs on a lily pad. 7 of the frogs hop in the water. How many frogs are left on the lily pad?

______ frogs

Test Practice

19. $9 - 1 =$ ______

6 ○ 7 ○ 8 ○ 9 ○

Math at Home Collect a group of 9 objects. Have your child show you all of the numbers they can subtract from 9. Have him or her write a subtraction number sentence.

Name

Subtract from 10

Lesson 12

ESSENTIAL QUESTION
How do you subtract numbers?

Explore and Explain

Tools

____ − ____ = ____

Write your subtraction sentence here.

Teacher Directions: Use [cube] to model. There are 10 tigers playing in a field. 4 tigers go away. How many tigers are left? Trace your cubes and mark Xs on the tigers that go away. Write the subtraction number sentence.

See and Show

Processes & Practices

There are many ways to subtract from 10.

Subtract 5 from 10.

10 − 5 = 5   The difference is 5.

Subtract 7 from 10.

10 − 7 = 3   The difference is 3.

Start with 10 . Subtract some cubes.
Write different ways to subtract from 10.

Subtract from 10

1. 10 − ____ = ____	2. 10 − ____ = ____
3. 10 − ____ = ____	4. 10 − ____ = ____
5. 10 − ____ = ____	6. 10 − ____ = ____
7. 10 − ____ = ____	8. 10 − ____ = ____

Talk Math When would you use subtraction in a real-world situation? Explain.

Name ....

On My Own

Use Work Mat 3 and [cube]. Subtract.

9. 10 − 3 = ______

10. 10 − 1 = ______

11. 9 − 5 = ______

12. 10 − 2 = ______

13. 10 − 0 = ______

14. 10 − 9 = ______

15. 6 − 3 = ______

16. 10 − 4 = ______

17. 10 − 8 = ______

18. 7 − 7 = ______

19. 10 − 7 = ______

20. 9 − 3 = ______

21. $\begin{array}{r} 10 \\ -\ 5 \\ \hline \end{array}$

22. $\begin{array}{r} 10 \\ -\ 2 \\ \hline \end{array}$

23. $\begin{array}{r} 4 \\ -\ 2 \\ \hline \end{array}$

24. $\begin{array}{r} 8 \\ -\ 3 \\ \hline \end{array}$

25. $\begin{array}{r} 10 \\ -\ 6 \\ \hline \end{array}$

26. $\begin{array}{r} 10 \\ -\ 7 \\ \hline \end{array}$

Problem Solving

Solve. Use Work Mat 3 and if needed.

27. Marisa sees 10 flamingos in a lake. 2 of them get out of the lake. How many flamingos are still in the lake?

_____ flamingos

28. There are 10 jeeps on a safari. 5 of the jeeps leave. How many jeeps are still on the safari?

_____ jeeps

HOT Problem There are 10 parrots on a branch. 7 of them fly away. The answer is 3 parrots. What is the question?

Name ....

My Homework

Lesson 12
Subtract from 10

Homework Helper

eHelp Need help? connectED.mcgraw-hill.com

You can subtract many different numbers from 10.

10 − 9 = 1   The difference is 1.

10 − 5 = 5   The difference is 5.

Practice

Write different ways to subtract from 10.

1. 10 − ______ = ______

2. 10 − ______ = ______

3. 10 − ______ = ______

4. 10 − ______ = ______

5. 10 − ______ = ______

6. 10 − ______ = ______

7. 10 − ______ = ______

8. 10 − ______ = ______

Subtract.

9. $9 - 6 =$ ______

10. $10 - 4 =$ ______

11. $10 - 8 =$ ______

12. $10 - 5 =$ ______

13. $\begin{array}{r} 10 \\ -\ 9 \\ \hline \end{array}$

14. $\begin{array}{r} 8 \\ -\ 2 \\ \hline \end{array}$

15. $\begin{array}{r} 10 \\ -\ 0 \\ \hline \end{array}$

16. Chris saw 10 flies sitting on an elephant. 6 of them flew away. How many flies are still sitting on the elephant?

______ flies

Test Practice

17. $10 - 2 =$ ______

6 ○ 7 ○ 8 ○ 9 ○

Math at Home Ask your child to use 10 buttons or pennies to show all of the ways they can subtract from 10. Have him or her write a subtraction number sentence to show one of the ways they subtracted from 10.

Name ....

Relate Addition and Subtraction

Lesson 13

ESSENTIAL QUESTION
How do you subtract numbers?

Explore and Explain

Watch Tools

Rub-a-dub-dub!
☐ + ☐ = ☐
☐ – ☐ = ☐
Write your number sentences here.

Teacher Directions: Use ●● to model. There are 3 birds in the bird bath. 2 more birds join them. Write the addition number sentence to show how many birds there are in all. Write a related subtraction number sentence.

See and Show

Processes & Practices

Related facts use the same numbers.
These facts can help you add and subtract.

$5 + 2 = 7$
$7 - 5 = 2$
$7 - 2 = 5$

Helpful Hint
$5 + 2 = 7$.
Use that fact to find
$7 - 2 = 5$.

You can use 5 + 2 = 7
to find 7 − 2 = 5.

They are opposite or inverse operations.

Identify an addition fact. Use Work Mat 3 and [counters] to find a related subtraction fact. Write both facts.

1.

Part	Part
3	6
Whole	
9	

____ + ____ = ____

____ − ____ = ____

2.

Part	Part
2	4
Whole	
6	

____ + ____ = ____

____ − ____ = ____

Talk Math How can addition facts help you subtract? Explain.

Name ______________________________

On My Own

Use Work Mat 3 and [counters] to find the related subtraction facts. Write the facts.

3. 5 + 4 = 9

___ − ___ = ___

___ − ___ = ___

4. 8 + 1 = 9

___ − ___ = ___

___ − ___ = ___

5. 1 + 4 = 5

___ − ___ = ___

___ − ___ = ___

6. 3 + 4 = 7

___ − ___ = ___

___ − ___ = ___

7. 5 + 3 = 8

___ − ___ = ___

___ − ___ = ___

8. 2 + 3 = 5

___ − ___ = ___

___ − ___ = ___

Real World! Problem Solving

Processes & Practices

Write a subtraction number sentence.
Then write a related addition fact.

9. There are 8 lizards. 6 of the lizards run away. How many lizards are left?

_____ − _____ = _____ lizards

_____ + _____ = _____

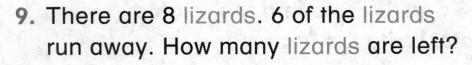

10. Chad saw 7 birds. Audrey saw 4 birds. How many more birds did Chad see?

_____ − _____ = _____ birds

_____ + _____ = _____

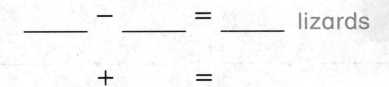

Write Math $5 + 4 = 9$ and $9 - 3 = 6$. Are these related facts? Explain why or why not.

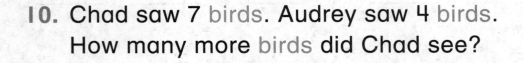

Name ....

My Homework

Lesson 13
Relate Addition and Subtraction

Homework Helper

eHelp Need help? connectED.mcgraw-hill.com

You can write related addition and subtraction facts. Related facts use the same numbers.

$2 + 3 = 5$
$5 - 2 = 3$
$5 - 3 = 2$

Helpful Hint
Use $2 + 3 = 5$ to find $5 - 2 = 3$ or $5 - 3 = 2$.

Practice

Identify an addition fact.
Write a related addition and subtraction fact.

1.

Part	Part
4	3
Whole	
7	

____ + ____ = ____

____ − ____ = ____

2.

Part	Part
3	5
Whole	
8	

____ + ____ = ____

____ − ____ = ____

Find the related subtraction facts. Write the facts.

3. $4 + 2 = 6$

____ − ____ = ____

____ − ____ = ____

4. $1 + 8 = 9$

____ − ____ = ____

____ − ____ = ____

Write the subtraction number sentence.
Then write the related addition fact.

5. There are 6 flowers in a vase. Mom gives 2 of the flowers to a friend. How many flowers are in the vase now?

____ − ____ = ____

____ + ____ = ____

Vocabulary Check

Vocab

Circle the correct answer.

subtract **related facts**

6. $3 + 1 = 4$ and $4 - 1 = 3$ are ________.

Math at Home Write an addition or subtraction number sentence using numbers 1–9. Have your child write a related addition or subtraction fact.

Name

True and False Statements

Lesson 14

ESSENTIAL QUESTION
How do you subtract numbers?

What's all the buzz about?

Explore and Explain

Tools

true
false

Teacher Directions: Use cubes to model. There are 6 bees making honey. 3 bees fly away. Trace your cubes. Mark Xs to show the bees that fly away. Liam says that 4 bees are still making honey. Is this true or false? Circle it.

See and Show

Processes & Practices

In math, statements can be true or false.
A true statement is correct.

9 − 4 = 5 is true.

A false statement is incorrect.

8 − 3 = 1 is false.

Determine if each subtraction statement is true or false. Circle true or false.

1. 8 − 4 = 5

   true   false

2. 9 − 0 = 0

   true   false

3. 5 − 3 = 2

   true   false

4. 7 − 4 = 3

   true   false

5. 8 − 7 = 1

   true   false

6. 0 = 9 − 0

   true   false

Talk Math How do you know when a subtraction statement is true? Explain.

Name ....

On My Own

Determine if each subtraction statement is true or false. Circle true or false.

7. $9 - 1 = 10$

true false

8. $5 - 2 = 2$

true false

9. $7 - 6 = 1$

true false

10. $8 - 0 = 0$

true false

11. $\begin{array}{r} 8 \\ -\ 2 \\ \hline 7 \end{array}$

true false

12. $\begin{array}{r} 4 \\ -\ 3 \\ \hline 1 \end{array}$

true false

13. $7 = 8 - 1$

true false

14. $2 = 6 - 3$

true false

15. $9 - 7 = 3$

true false

16. $7 - 1 = 6$

true false

Problem Solving

Determine if the word problem is true or false. Circle true or false.

17. 6 beetles are crawling on the ground. 2 fly away. There are 4 beetles left.

true false

18. There are 7 spiders on a web. 5 of those spiders crawl away. There are 3 spiders left on the web.

true false

Write Math Write your own false statement. Explain why your statement is false.

Name ....

My Homework

Lesson 14

True and False Statements

Homework Helper

eHelp Need help? connectED.mcgraw-hill.com

A true math statement is correct.
A false math statement is incorrect.

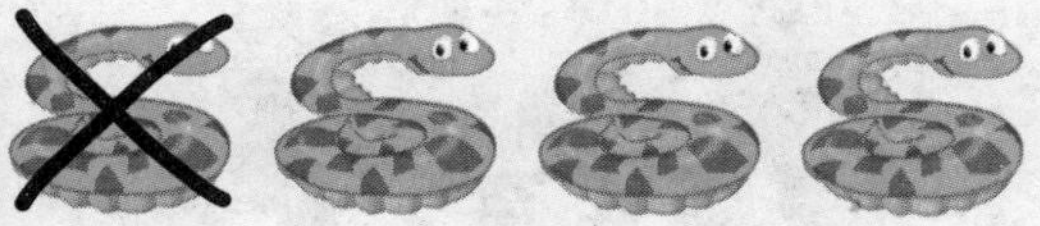

4 − 1 = 3

true (circled) false

6 − 3 = 4

true false (circled)

Practice

Determine if each subtraction statement is true or false. Circle true or false.

1. 8 − 8 = 8

true false

2. 4 − 1 = 5

true false

3. 6 = 9 − 3

true false

4. 7 − 1 = 6

true false

Determine if each subtraction statement is true or false. Circle true or false.

5.

$$\begin{array}{r} 6 \\ -\ 3 \\ \hline 3 \end{array}$$

true false

6.

$$\begin{array}{r} 9 \\ -\ 1 \\ \hline 9 \end{array}$$

true false

7. 5 birds are sitting on a branch. 1 bird flies away. There are 6 birds left on the branch.

true false

8. There are 8 gorillas walking together. 3 of them walk away. 5 gorillas are still walking together.

true false

Test Practice

9. Which subtraction statement is true?

$5 - 1 = 3$	$5 + 2 = 8$	$7 - 3 = 2$	$9 - 6 = 3$
○	○	○	○

Math at Home Write a false subtraction sentence using numbers 1–9. Ask your child if the problem is true or false. Have your child correct the number sentence.

Name ______________________

Processes & Practices

Fluency Practice

Subtract.

1. 5 − 3 = ______
2. 10 − 4 = ______
3. 2 − 0 = ______
4. 6 − 3 = ______
5. 1 − 1 = ______
6. 9 − 5 = ______
7. 10 − 9 = ______
8. 7 − 3 = ______
9. 4 − 1 = ______
10. 6 − 5 = ______
11. 9 − 3 = ______
12. 8 − 0 = ______
13. 5 − 1 = ______
14. 10 − 3 = ______
15. 9 − 2 = ______
16. 4 − 4 = ______
17. 2 − 2 = ______
18. 5 − 2 = ______
19. 8 − 4 = ______
20. 7 − 6 = ______
21. 6 − 2 = ______
22. 9 − 1 = ______
23. 7 − 2 = ______
24. 4 − 2 = ______

Fluency Practice

Subtract.

1. $8 - 3$	2. $10 - 4$	3. $9 - 6$	4. $5 - 5$
5. $4 - 2$	6. $6 - 1$	7. $9 - 0$	8. $3 - 2$
9. $10 - 5$	10. $7 - 3$	11. $1 - 1$	12. $8 - 6$
13. $2 - 0$	14. $9 - 8$	15. $7 - 2$	16. $5 - 3$
17. $8 - 4$	18. $6 - 5$	19. $3 - 3$	20. $10 - 1$
21. $7 - 5$	22. $9 - 3$	23. $10 - 3$	24. $4 - 1$

Name ....

My Review

Chapter 2

Subtraction Concepts

Vocabulary Check

Vocab

Complete each sentence.

difference **related facts**

subtract **subtraction number sentence**

1. Addition and subtraction number sentences that use the same numbers are called ______________.

2. 9 − 7 = 2 is a ______________ ______________.

3. When you take away, you ______________.

4. In 7 − 4 = 3, the 3 is the ______________.

Concepts Check

Write a subtraction number sentence.

5.

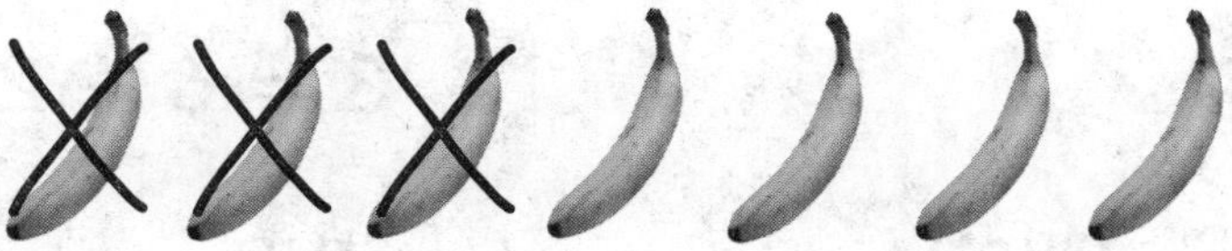

____ ○ ____ ○ ____

Subtract.

6. 5 − 1 = _____

7. 4 − 2 = _____

8. 7 − 5 = _____

9. 9 − 5 = _____

10. $\begin{array}{r} 8 \\ -\ 4 \\ \hline \end{array}$

11. $\begin{array}{r} 6 \\ -\ 1 \\ \hline \end{array}$

12. $\begin{array}{r} 4 \\ -\ 0 \\ \hline \end{array}$

13. $\begin{array}{r} 9 \\ -\ 2 \\ \hline \end{array}$

14. $\begin{array}{r} 8 \\ -\ 3 \\ \hline \end{array}$

15. $\begin{array}{r} 7 \\ -\ 5 \\ \hline \end{array}$

Write the related subtraction facts.

16. 6 + 2 = _____

_____ − _____ = _____

_____ − _____ = _____

Determine if each subtraction statement is true or false. Circle true or false.

17. 10 − 3 = 7

true   false

18. 3 = 6 − 2

true   false

Name ....

Real World Problem Solving

Write a subtraction number sentence.

19. Angel has 7 bananas. She eats 2 of them. How many bananas are left?

☐
− ☐
☐ bananas

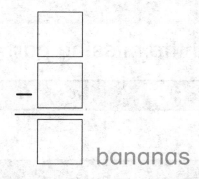

Circle true or false.

20. There are 9 bears on a tree. 3 of them leave. 6 bears are still on the tree.

true   false

Test Practice

Find the matching subtraction number sentence.

21. There are 5 elephants. There are 3 leopards. How many fewer leopards are there?

$5 - 3 = 1$ fewer leopards ○

$5 - 3 = 2$ fewer leopards ○

$5 - 3 = 1$ fewer elephants ○

$5 - 3 = 4$ fewer elephants ○

Reflect

Chapter 2
Answering the Essential Question

Show ways to answer.

Find the missing part.

Part	Part
____	3
Whole	
7	

Subtract 0.

8 − 0 = ____

ESSENTIAL QUESTION
How do you subtract numbers?

Subtract.

$$\begin{array}{r} 9 \\ -\ 7 \\ \hline \square \end{array}$$

Write a subtraction number sentence.

____ − ____ = ____

Ready?
Set?
Solve!

Chapter

3 Addition Strategies to 20

ESSENTIAL QUESTION

How do I use strategies to add numbers?

We're in the Big City!

Watch a video!

Name

Chapter 3 Project

My Addition Story

1. Design a cover for your addition story book below. Title the book *My Addition Story Book.*
2. Write an addition story on each page. Include an addition number sentence and a drawing that matches each story.
3. Write at least 4 addition stories in all.

Name ....

Am I Ready?

1. Circle the addition sign.

   \+   −   =

2. Circle the equals sign.

   \+   −   =

Add.

3. $\begin{array}{r} 3 \\ +\ 0 \\ \hline \end{array}$

4. $\begin{array}{r} 2 \\ +\ 2 \\ \hline \end{array}$

5. $\begin{array}{r} 7 \\ +\ 1 \\ \hline \end{array}$

6. $\begin{array}{r} 6 \\ +\ 3 \\ \hline \end{array}$

7. $\begin{array}{r} 4 \\ +\ 1 \\ \hline \end{array}$

8. $\begin{array}{r} 3 \\ +\ 4 \\ \hline \end{array}$

Use the pictures to write an addition number sentence.

9. 

_____ ◯ _____ ◯ _____

How Did I Do?

Shade the boxes to show the problems you answered correctly.

1	2	3	4	5	6	7	8	9

Name

My Math Words

Vocab

Review Vocabulary

minus (−)   plus (+)

Use the review words to fill in the blanks below. Then write an addition and a subtraction number sentence.

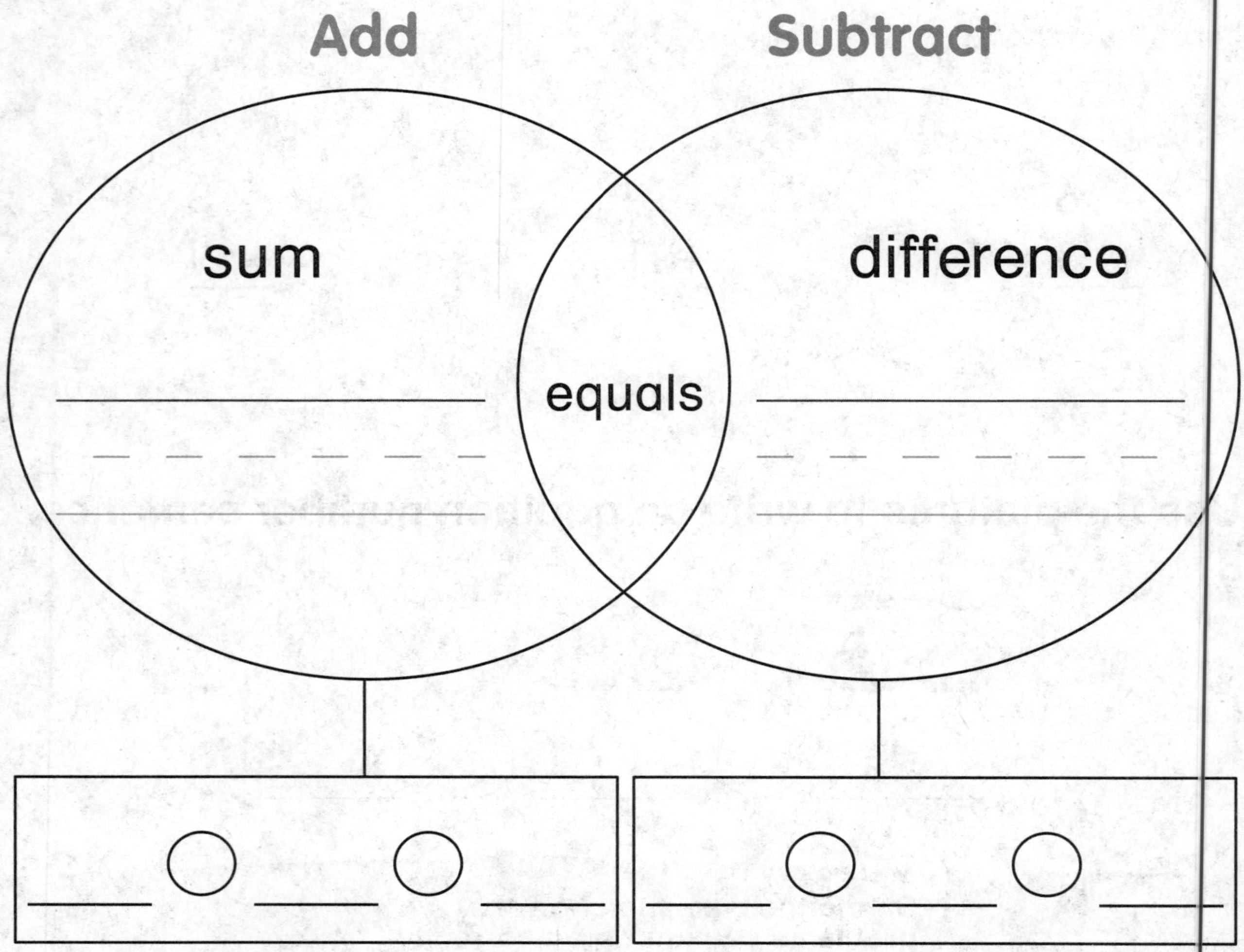

My Vocabulary Cards

Vocab

Processes & Practices

Lesson 3-4

addends

8 + 9 = 17

addends

Lesson 3-1

count on

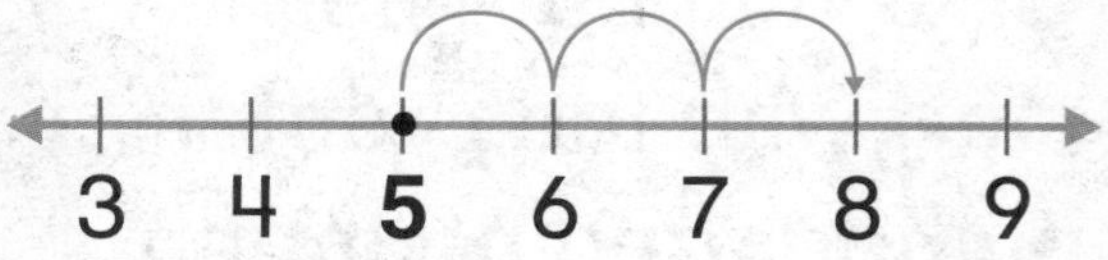

5 + 3 = 8

Lesson 3-4

doubles

3 + 3 = 6

Lesson 3-5

doubles minus 1

3 + 2 = 5

Lesson 3-5

doubles plus 1

3 + 4 = 7

Lesson 3-3

number line

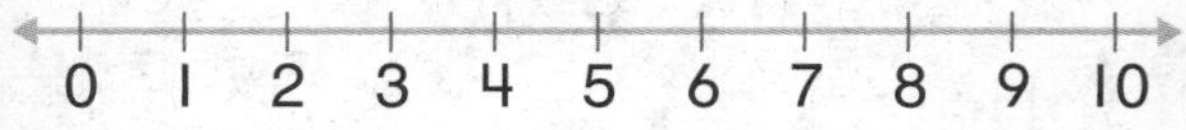

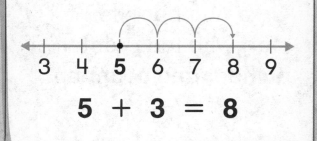

Teacher Directions:
Ideas for Use

- Have students arrange the cards in alphabetical order.
- Ask students to write a tally mark on each card every time they read or hear the word.

On a number line, start with the greater number and count up.

Any numbers being added together.

Add with doubles and subtract one.

Two addends that are the same.

A line with number labels.

Add with doubles and add one.

My Foldable

FOLDABLES® Follow the steps on the back to make your Foldable.

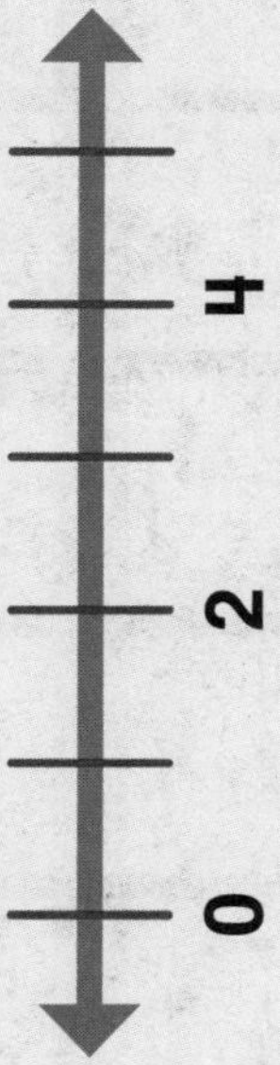

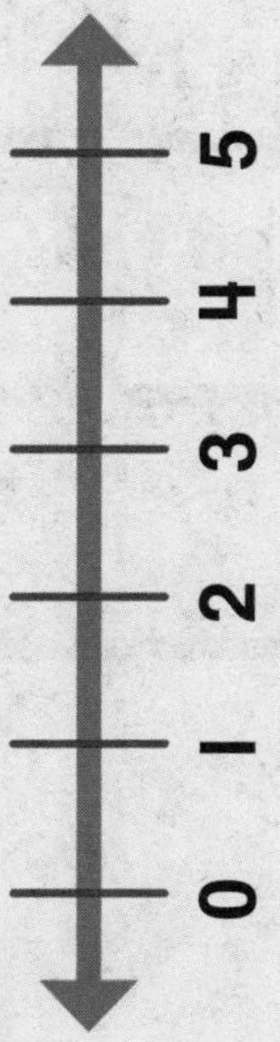

FOLDABLES® Study Organizer

1

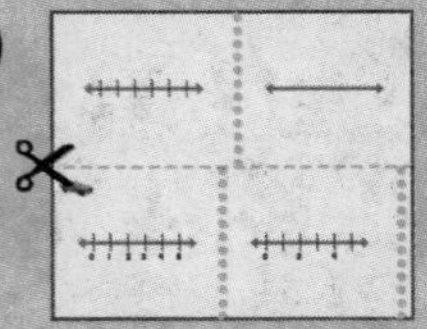

2

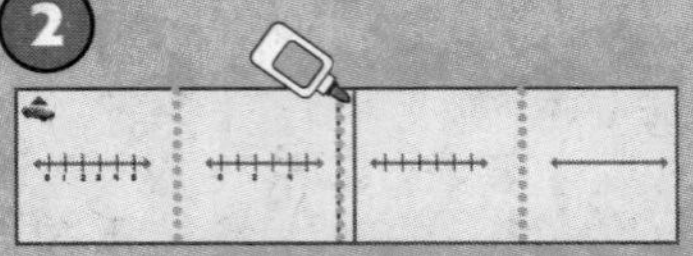

3

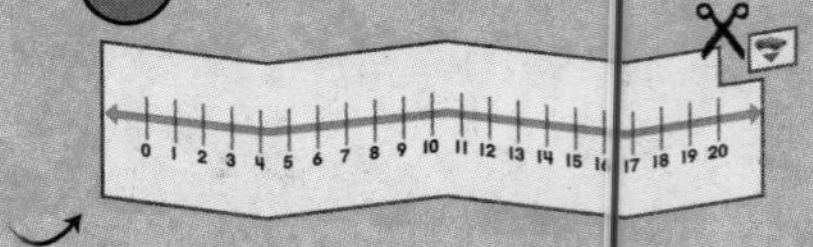

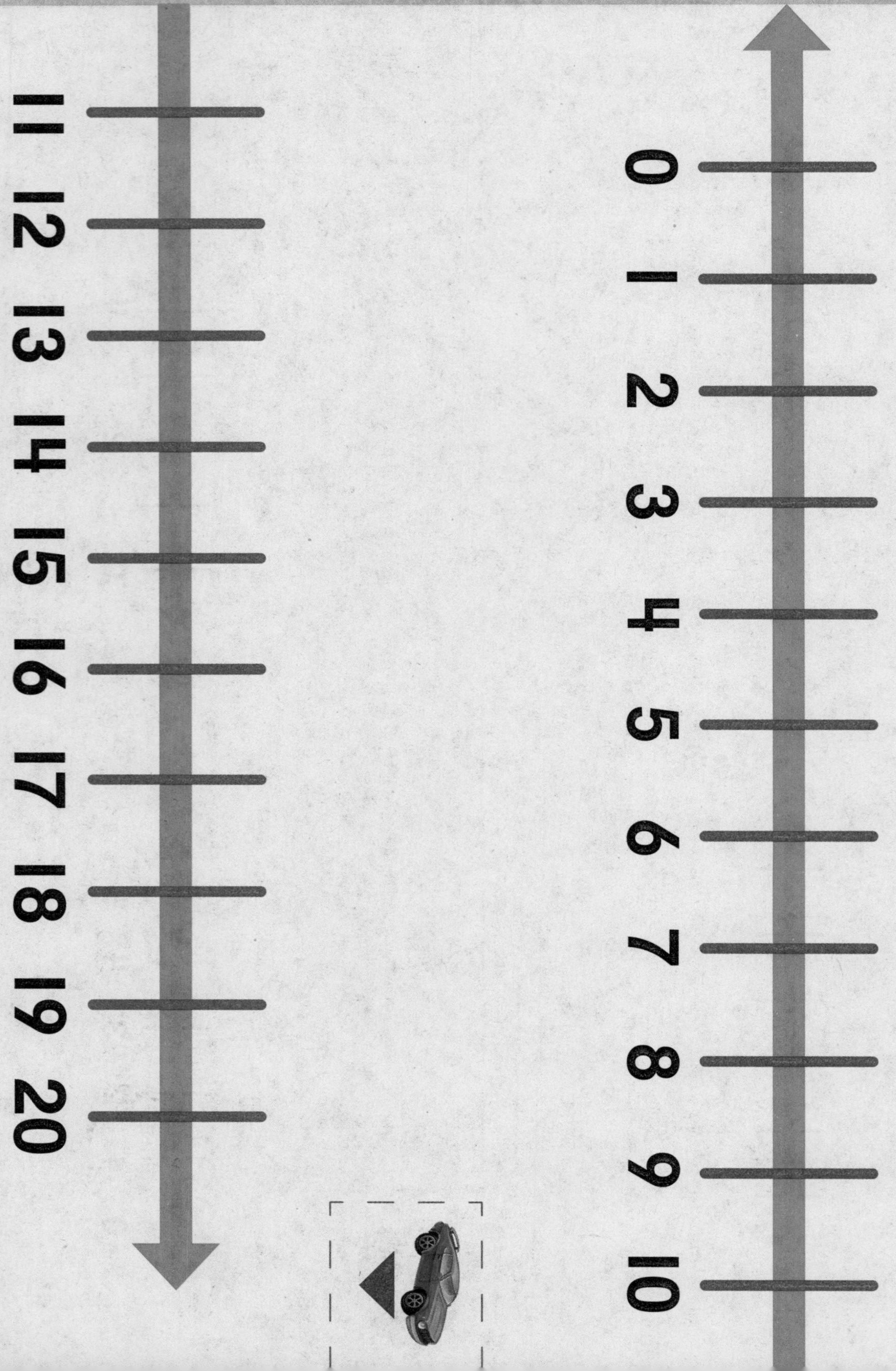

Name ....

Count On 1, 2, or 3

Lesson 1

ESSENTIAL QUESTION
How do I use strategies to add numbers?

Explore and Explain

Watch Tools

_____ + _____ = _____

Write your addition sentence here.

Teacher Directions: Draw a group of four crayons. Then draw 1, 2, or 3 more crayons. Write an addition number sentence that tells how many in all.

See and Show

Processes & Practices

You can **count on** to add. There are 5 crayons in the group. Add 2 more crayons.

5, 6, 7

5 + 2 = 7

Helpful Hint
Start with the greater number, 5.
Count on 2 more.
5, 6, 7
5 + 2 = 7

Use . Start with the greater number. Count on to add.

1.

Count on 3 is to add 3.

7, ______, ______, ______

7 + 3 = ______

2.

Count on 2 is to add 2.

6, ______, ______

2 + 6 = ______

Talk Math Tell how to count on to add 5 + 3.

Name ....

On My Own

Start with the greater number. Count on to add.

3. 5 + 3 = ______

4. 8 + 3 = ______

5. 4 + 1 = ______

6. 1 + 2 = ______

7. 9 + 3 = ______

8. 1 + 8 = ______

9. 3 + 7 = ______

10. 2 + 9 = ______

11. 1 + 7 = ______

12. 4 + 3 = ______

13. 2 + 7 = ______

14. 5 + 1 = ______

15. $\begin{array}{r} 8 \\ +\ 2 \\ \hline \end{array}$

16. $\begin{array}{r} 1 \\ +\ 6 \\ \hline \end{array}$

17. $\begin{array}{r} 9 \\ +\ 1 \\ \hline \end{array}$

18. $\begin{array}{r} 3 \\ +\ 6 \\ \hline \end{array}$

19. $\begin{array}{r} 2 \\ +\ 5 \\ \hline \end{array}$

20. $\begin{array}{r} 3 \\ +\ 2 \\ \hline \end{array}$

Problem Solving

Processes & Practices

21. Bella sees 2 buses. Then she sees 3 more buses. How many buses does she see in all?

_______ buses

22. Jake saw 5 cars in the morning. He saw 3 more cars in the evening. How many cars did Jake see in all?

_______ cars

Write Math Explain how you count on to find $3 + 7$.

Name ____________________

My Homework

Lesson 1

Count On 1, 2, or 3

Homework Helper

eHelp Need help? connectED.mcgraw-hill.com

You can count on to add.

Helpful Hint
To count on, start with the greater number.

4, 5, 6, 7

4 + 3 = 7

Practice

Start with the greater number. Count on to add.

1. 7 + 2 = ______
2. 4 + 3 = ______
3. 5 + 3 = ______
4. 6 + 1 = ______
5. 9 + 1 = ______
6. 2 + 5 = ______
7. 8 + 3 = ______
8. 7 + 1 = ______

Start with the greater number. Count on to add.

9. $3 + 6$

10. $8 + 2$

11. $1 + 8$

12. $3 + 9$

13. $6 + 2$

14. $9 + 2$

15. Drew went to basketball practice 3 times this week. He went to soccer practice 2 times this week. How many practices did he go to in all?

Practice makes perfect!

_____ practices

Vocabulary Check

Vocab

Circle the missing word.

count on **sum**

16. You can _________ to add when you join any number with 1, 2, or 3.

Math at Home Say a number between 1 and 9. Ask your child to add 1, 2, and 3 to that number.

Name ______________________

Count On Using Pennies

Lesson 2

ESSENTIAL QUESTION
How do I use strategies to add numbers?

Explore and Explain

Tools

_____ pennies

Teacher Directions: Use [penny] to model. Reese has 6 pennies in his bank. He puts in 3 more pennies. Count on to find how many pennies Reese has in all. Write how many there are in all. Trace the pennies to show your work.

See and Show

Processes & Practices

A penny has a value of 1 cent.
You can count on by ones to add pennies.

penny

1 cent = 1¢

Helpful Hint
Start with 8 pennies. Count on 9, 10, 11. There are 11 pennies in all.

8¢, 9 ¢, 10 ¢, 11 ¢

Count the group of pennies. Then count on to add.

1.

Count on 2 is to add 2.

7¢, ____¢, ____¢

Talk Math Why do you count on by ones when you use pennies?

Name ....

On My Own

Count the group of pennies.
Then count on to add.

2. 

Count on 3.

4¢, _____ ¢, _____ ¢, _____ ¢

3. 

Count on 2.

10¢, _____ ¢, _____ ¢

4. 

Count on 1.

8¢, _____ ¢

Problem Solving

Real World

Processes & Practices

5. Kiah has 6 pennies. She is given 3 more. How many pennies does Kiah have now?

Does this make sense to you?

______ pennies

6. Enrique has 9 pennies in his left pocket. He has 2 more pennies in his right pocket. How many pennies does he have in all?

______ pennies

HOT Problem Eliana buys an eraser for 10 pennies. She buys a sticker for 2 pennies. The answer is 12 pennies. What is the question?

Name ....

My Homework

Lesson 2
Count On Using Pennies

Homework Helper

eHelp Need help? connectED.mcgraw-hill.com

You can count on by ones to add pennies.

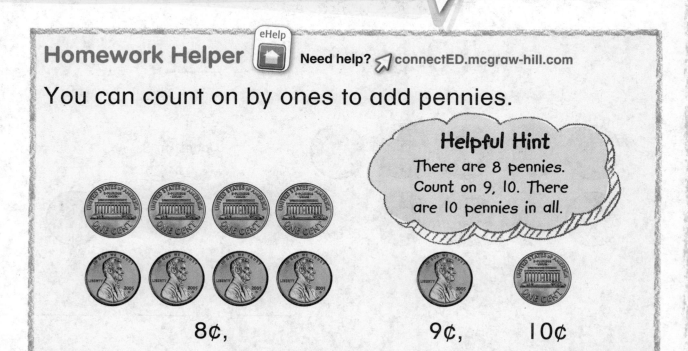

Helpful Hint
There are 8 pennies. Count on 9, 10. There are 10 pennies in all.

8¢,

9¢, 10¢

Practice

Count the group of pennies. Then count on to add.

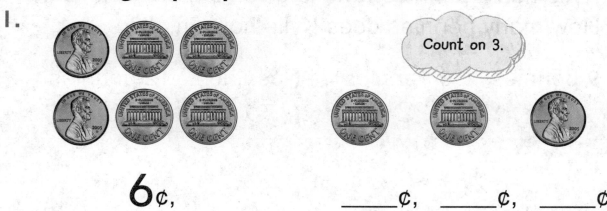

1.

Count on 3.

6¢, ____¢, ____¢, ____¢

Count the group of pennies. Then count on to add.

2.

10¢, ____¢

3.

9 ¢, ____¢, ____¢, ____¢

Test Practice

4. Kylie has 7 pennies. She is given 3 more pennies. How many pennies does Kylie have in all?

9 pennies	10 pennies	11 pennies	12 pennies
○	○	○	○

Math at Home Give your child 10 pennies. Provide your child with 2 more pennies. Ask him or her to count on by 2 using those pennies. Have your child tell you how many pennies he or she has in all.

Name ....

Use a Number Line to Add

Lesson 3

ESSENTIAL QUESTION
How do I use strategies to add numbers?

Off to the city!

Explore and Explain

Watch Tools

_____ + _____ = _____

Write your addition sentence here.

Teacher Directions: Use [cube] to model. A car is driving across the bridge. It starts at the number 4. It drives 3 spaces to the right. Where does the car stop? Write the addition number sentence.

See and Show

Processes
&Practices

You can use a **number line** to add.
Start with the greater number and count on by moving to the right.

Helpful Hint
Start with the greater number. 5, 6, 7, 8

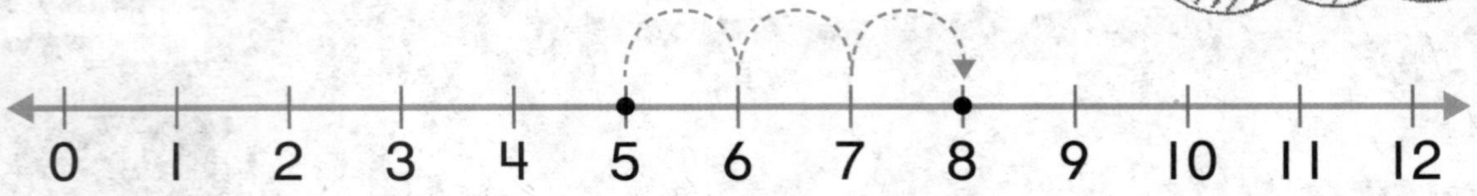

5 + 3 = 8

Use the number line to add. Show your work. Write the sum.

1. 6 + 2 = ____

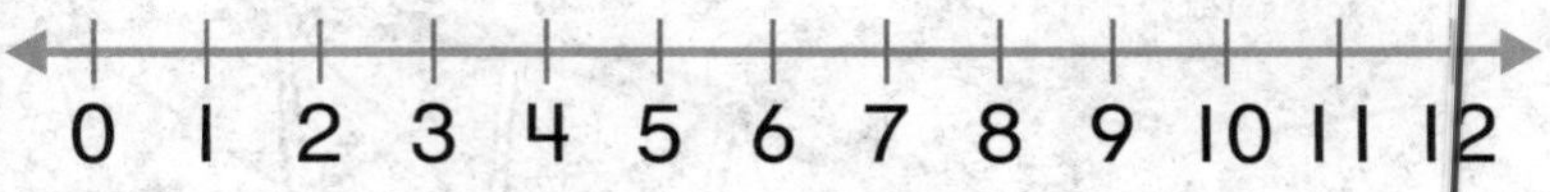

2. 8 + 2 = ____

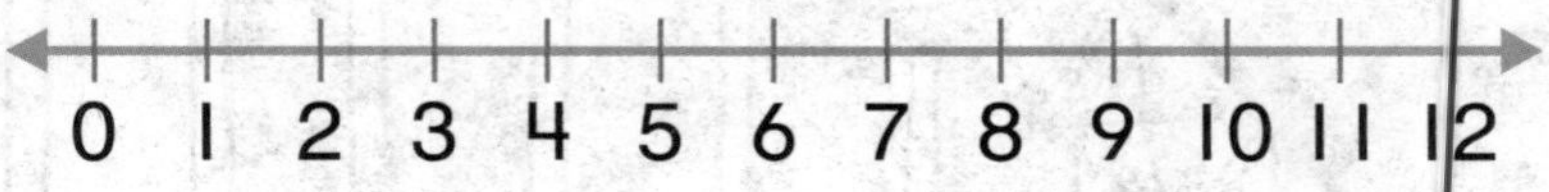

3. 1 + 4 = ____

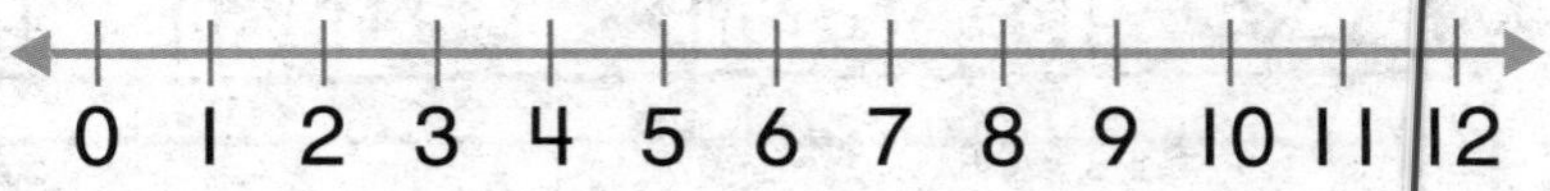

4. 7 + 2 = ____

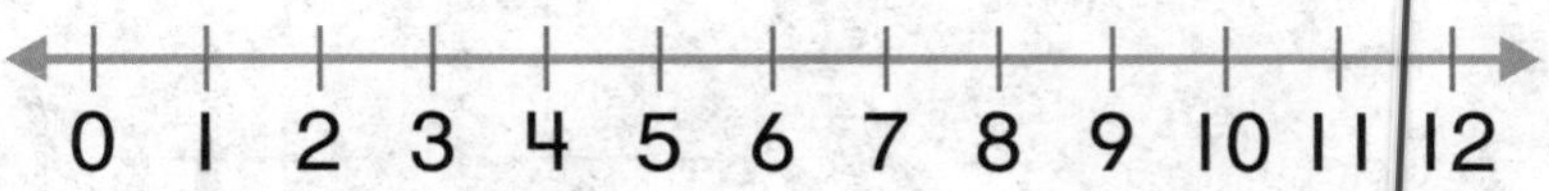

Talk Math How does a number line help you add?

Name ....

On My Own

Use the number line to add. Write the sum.

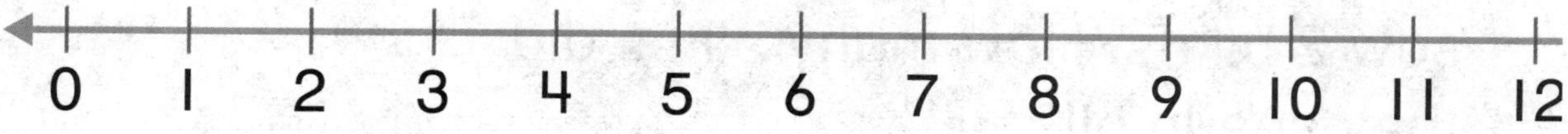

5. 3 + 4 = ______

6. 2 + 9 = ______

7. 1 + 8 = ______

8. 7 + 3 = ______

9. $\begin{array}{r} 6 \\ +\ 1 \\ \hline \end{array}$

10. $\begin{array}{r} 8 \\ +\ 3 \\ \hline \end{array}$

11. $\begin{array}{r} 1 \\ +\ 9 \\ \hline \end{array}$

12. $\begin{array}{r} 2 \\ +\ 7 \\ \hline \end{array}$

13. $\begin{array}{r} 9 \\ +\ 3 \\ \hline \end{array}$

14. $\begin{array}{r} 5 \\ +\ 2 \\ \hline \end{array}$

15. $\begin{array}{r} 1 \\ +\ 7 \\ \hline \end{array}$

16. $\begin{array}{r} 4 \\ +\ 2 \\ \hline \end{array}$

17. $\begin{array}{r} 6 \\ +\ 3 \\ \hline \end{array}$

Problem Solving

Processes & Practices

Use the number line to solve.

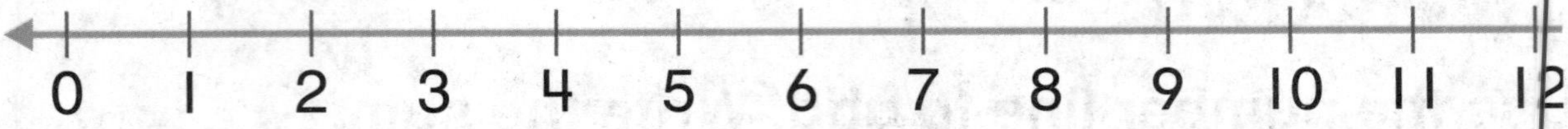

18. Amelio saw 5 bikes. His brother saw 2 bikes. How many bikes did they see in all?

_______ bikes

19. Liz saw 8 pigeons. Logan saw 2 pigeons. How many pigeons did they see in all?

_______ pigeons

Write Math Why do you start with the greater number when adding on a number line?

Name ______________________

My Homework

Lesson 3
Use a Number Line to Add

Homework Helper

eHelp Need help? connectED.mcgraw-hill.com

You can use a number line to add.

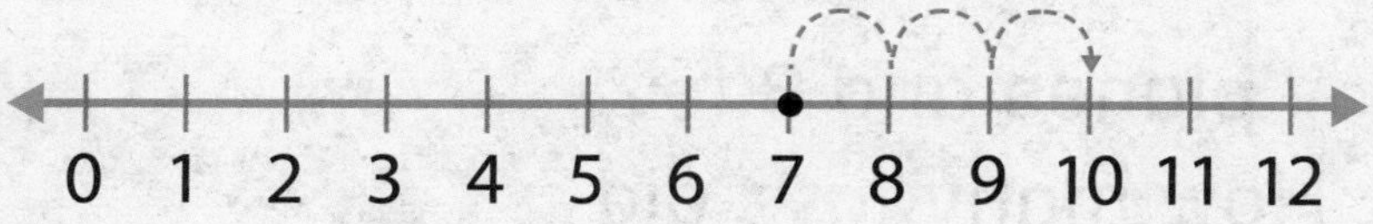

7 + 3 = 10

Helpful Hint
Start with the greater number and count on by moving to the right.

Practice

Use the number line above to add. Write the sum.

1. 1 + 8 = ______

2. 8 + 2 = ______

3. 5 + 3 = ______

4. 8 + 3 = ______

5. 7 + 2 = ______

6. 4 + 1 = ______

7.
```
  9
+ 2
---
```

8.
```
  4
+ 3
---
```

9.
```
  6
+ 3
---
```

Use the number line to solve.

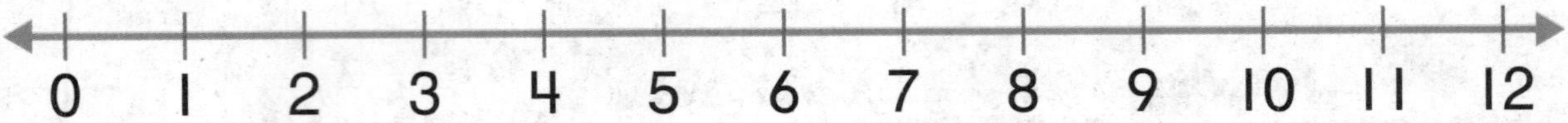

10. There are 6 people waiting for a taxi. 2 more people get in line. How many people are waiting for a taxi in all?

_____ people

11. Abigail puts 9 toy planes and 3 toy cars on a shelf. How many toys did she put on the shelf?

_____ toys

Vocabulary Check

Vocab

Complete each sentence.

number line **count on**

12. When you add on a _______________, start with the greater number and move to the right.

13. To _______________, start with the greater number.

Math at Home Help your child create a number line from 0 to 10. Then, ask your child to use the number line to show 1 + 9.

Name ....

Use Doubles to Add

Lesson 4

ESSENTIAL QUESTION
How do I use strategies to add numbers?

Explore and Explain

Watch Tools

_____ + _____ = _____

Write your addition sentence here.

Teacher Directions: Use [cube] to model. Build a tower that has 3 red cubes and 3 green cubes. Trace your tower. Write the addition number sentence.

See and Show

Processes &Practices

Addends are the numbers you add.
Both addends are the same in a **doubles** fact.

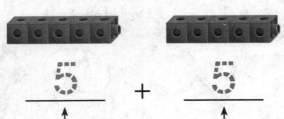

5 + 5 = 10

addend, addend, sum

Helpful Hint
5 + 5 is a doubles fact.

Use  to model. Complete the addition number sentence.

1. 

____ + ____ = ____

2. 

____ + ____ = ____

Add. Circle the doubles facts.

3. $\begin{array}{r} 7 \\ +\ 7 \\ \hline \end{array}$

4. $\begin{array}{r} 5 \\ +\ 5 \\ \hline \end{array}$

5. $\begin{array}{r} 7 \\ +\ 4 \\ \hline \end{array}$

6. 6 + 6 = ____

7. 9 + 9 = ____

Talk Math Can you use doubles to make a sum of 7?

Name ______________________

On My Own

Use  to model. Complete the addition number sentence.

8. 

_____ + _____ = _____

9. 

_____ + _____ = _____

10. 

_____ + _____ = _____

11. 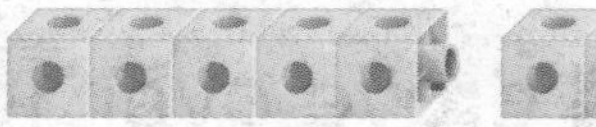

_____ + _____ = _____

Add. Circle the doubles facts.

12. $\begin{array}{r} 8 \\ +\ 8 \\ \hline \end{array}$

13. $\begin{array}{r} 9 \\ +\ 0 \\ \hline \end{array}$

14. $\begin{array}{r} 9 \\ +\ 9 \\ \hline \end{array}$

15. 8 + 3 = _____

16. 1 + 5 = _____

17. 6 + 6 = _____

18. 10 + 10 = _____

19. 3 + 3 = _____

20. 7 + 1 = _____

Problem Solving

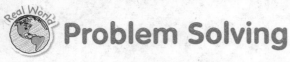

Processes & Practices

21. 4 taxi cabs drove down the road. 4 more taxi cabs drove down the road. How many taxi cabs drove down the road in all?

_____ taxi cabs

22. Emad drove over 1 bridge on his way to school. He drove over 1 more bridge on his way home. How many bridges did Emad drive over in all?

_____ bridges

Write Math Is 3 + 6 a doubles fact? Explain.

Name ....

My Homework

Lesson 4

Use Doubles to Add

Homework Helper

eHelp

Need help? connectED.mcgraw-hill.com

In a doubles fact, both addends are the same.

$$3 + 3 = 6$$

3 ↑ addend, 3 ↑ addend, 6 ↑ sum

Helpful Hint
3 + 3 is a doubles fact.

Practice

Add. Circle the doubles facts.

1. $2 + 2 =$ ____

2. $4 + 4 =$ ____

3. $2 + 9 =$ ____

4. $1 + 6 =$ ____

5. $5 + 5 =$ ____

6. $1 + 1 =$ ____

7. $7 + 7 =$ ____

8. $7 + 2 =$ ____

Add. Circle the doubles facts.

9. 6 + 3 = ______

10. 4 + 4 = ______

11. There are 4 tan cats sitting on a fence.
4 black cats are also on the fence.
How many cats are on the fence in all?

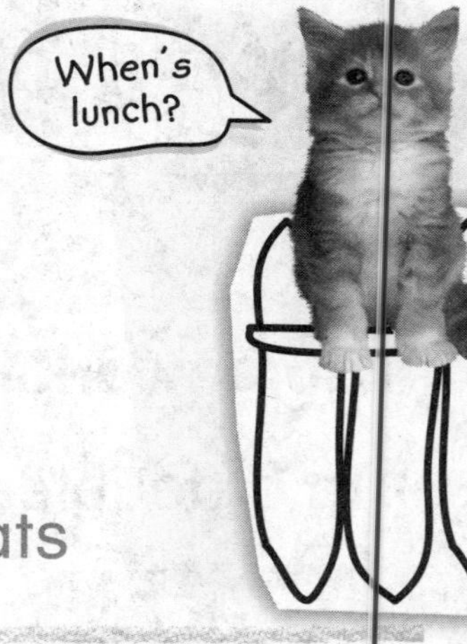

______ cats

12. There are 5 girls playing hopscotch. The same number of boys are playing catch.
How many children are playing in all?

______ children

Vocabulary Check

Vocab

Complete each sentence.

doubles **addends**

13. Numbers you add together to find a sum are called ______________.

14. Two addends that are the same number are ______________.

Math at Home Have your child identify things that show doubles such as fingers on both hands, toes on both feet, or windows in a car.

Name

Use Near Doubles to Add

Lesson 5

ESSENTIAL QUESTION
How do I use strategies to add numbers?

That's a bright idea!

Explore and Explain

Watch Tools

___ + ___ = ___
Write your addition sentence here.

Teacher Directions: Use cubes to model. Show the doubles fact 3 + 3 on the billboard. Add or take away one cube from one of the groups of cubes. Trace the cubes. Write the addition number sentence.

See and Show

Processes
&Practices

You can use near doubles facts to find a sum. If you know 5 + 5 = 10, you can find 5 + 6 and 5 + 4.

doubles	doubles plus 1	doubles minus 1

5 + 5 = 10	5 + 6 = 11	5 + 4 = 9

Use  to model. Write the addition number sentence.

1. 

_____ + _____ = _____

_____ + _____ = _____

2. 

_____ + _____ = _____

_____ + _____ = _____

Talk Math How do doubles facts help you learn near doubles facts?

Name ....

On My Own

Use to model. Find each sum.

3. $2 + 2 =$ ______

   $2 + 3 =$ ______

4. $3 + 3 =$ ______

   $3 + 2 =$ ______

5. $8 + 7 =$ ______

6. $7 + 7 =$ ______

7. $6 + 6 =$ ______

8. $6 + 5 =$ ______

9. $\begin{array}{r} 7 \\ +\ 6 \\ \hline \end{array}$

10. $\begin{array}{r} 7 \\ +\ 8 \\ \hline \end{array}$

11. $\begin{array}{r} 4 \\ +\ 3 \\ \hline \end{array}$

12. $\begin{array}{r} 5 \\ +\ 4 \\ \hline \end{array}$

13. $\begin{array}{r} 5 \\ +\ 5 \\ \hline \end{array}$

14. $\begin{array}{r} 5 \\ +\ 6 \\ \hline \end{array}$

15. $\begin{array}{r} 7 \\ +\ 3 \\ \hline \end{array}$

16. $\begin{array}{r} 7 \\ +\ 5 \\ \hline \end{array}$

17. $\begin{array}{r} 7 \\ +\ 4 \\ \hline \end{array}$

Problem Solving

Processes &Practices

Solve. Write the doubles fact that helped you solve the problem.

18. Tyra sees 5 beetles. Sara sees 6 beetles. How many beetles do they see in all?

_____ beetles

_____ + _____ = _____

19. Adam has 9 red flowers and 8 purple flowers. How many flowers does he have in all?

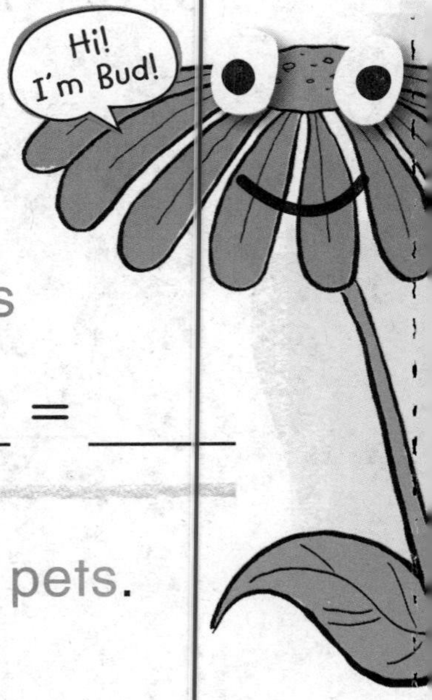

_____ flowers

_____ + _____ = _____

HOT Problem Devon has 6 pets. Maya has 7 pets. The answer is 13 pets. What is the question?

Name ______________________________

My Homework

Lesson 5

Use Near Doubles to Add

Homework Helper

eHelp

Need help? connectED.mcgraw-hill.com

You can use near doubles facts to find a sum.

doubles

3 + 3 = 6

doubles plus 1

3 + 4 = 7

doubles minus 1

3 + 2 = 5

Practice

Find each sum.

1. 4 + 4 = ____

4 + 5 = ____

2. 8 + 8 = ____

8 + 7 = ____

3. 6 + 5 = ____

4. 6 + 6 = ____

5. $\begin{array}{r} 7 \\ +\ 6 \\ \hline \end{array}$

6. $\begin{array}{r} 7 \\ +\ 7 \\ \hline \end{array}$

7. $\begin{array}{r} 7 \\ +\ 8 \\ \hline \end{array}$

Solve. Write the doubles fact that helped you solve the problem.

8. Paul is a dog walker. He walks 7 small dogs. He walks 6 large dogs. How many dogs does he walk in all?

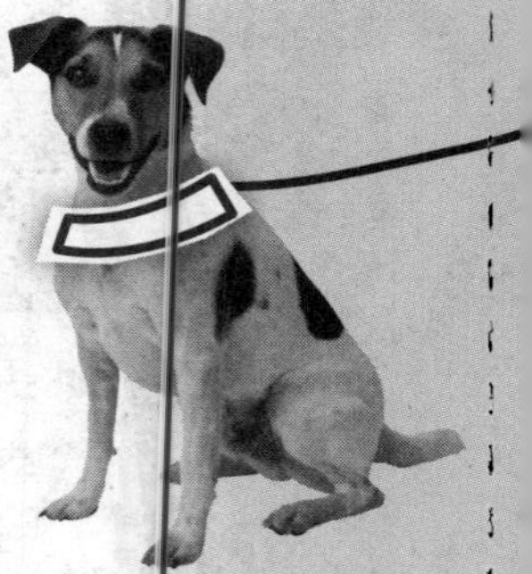

_____ dogs

_____ + _____ = _____

9. There are 9 girls and 8 boys in Maria's class who walk to school. How many children in her class walk to school in all?

_____ children

_____ + _____ = _____

Vocabulary Check

Vocab

Circle the missing word.

doubles plus 1 **doubles minus 1**

10. If you know that $8 + 8 = 16$, then you can use _______ to find $8 + 7$.

Math at Home Give your child an addition problem such as $4 + 5$ or $3 + 4$. Have your child give you the doubles fact that will help him or her find the sum.

Name ....

Check My Progress

Vocabulary Check

Draw lines to match.

1. count on	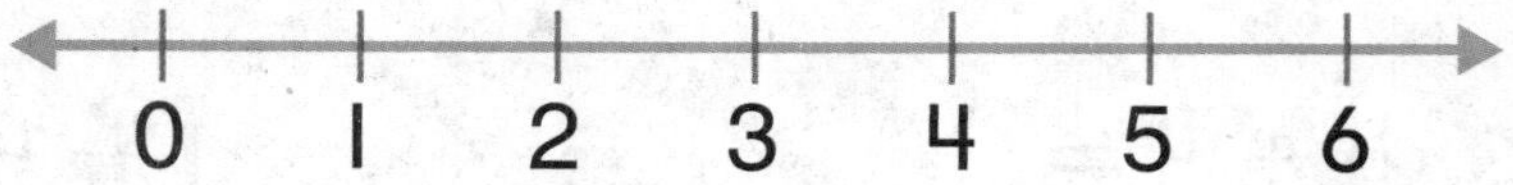
2. number line	4 + 4 = 8
3. addends	Start at a number and count forward to add.
4. doubles	Numbers you add together to find a sum.
5. doubles + 1	Add with doubles and subtract one.
6. doubles − 1	Add with doubles and add one.

Concept Check

Start with the greater number. Count on to add.

7. $\begin{array}{r} 6 \\ +\ 3 \\ \hline \end{array}$

8. $\begin{array}{r} 1 \\ +\ 7 \\ \hline \end{array}$

9. $\begin{array}{r} 2 \\ +\ 3 \\ \hline \end{array}$

Use the number line to add. Write the sum.

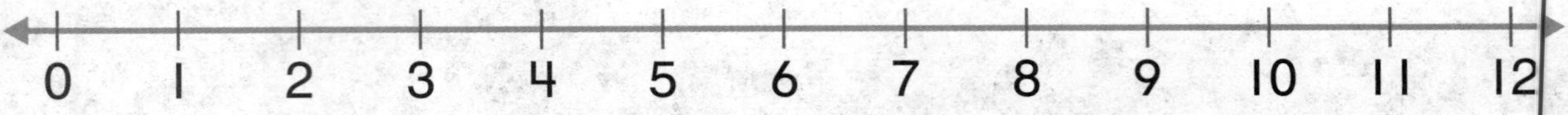

10. $\begin{array}{r} 3 \\ +\ 8 \\ \hline \end{array}$

11. $\begin{array}{r} 4 \\ +\ 2 \\ \hline \end{array}$

12. $\begin{array}{r} 7 \\ +\ 3 \\ \hline \end{array}$

Add. Circle the doubles facts.

13. 4 + 4 = ____

14. 4 + 5 = ____

15. 9 + 8 = ____

16. 8 + 8 = ____

17. Craig counted 7 doors. He counted 2 more doors. How many doors did Craig count in all?

____ doors

Test Practice

18. Which doubles fact helps you find this sum?

9 + 8 = ____

10 + 10	7 + 7	5 + 5	9 + 9
○	○	○	○

Name ....

Problem Solving

STRATEGY: Act It Out

Lesson 6

ESSENTIAL QUESTION
How do I use strategies to add numbers?

Watch Tools

3 red birds are on a branch.
There are 2 more yellow birds than red birds on another branch.
How many yellow birds are there?

1 Understand

Underline what you know.
Circle what you need to find.

2 Plan

How will I solve the problem?

3 Solve

I will act it out.

5 yellow birds

4 Check

Is my answer reasonable? Explain.

Practice the Strategy

Susan found 3 shells on the beach. Jamar found 1 more shell than Susan did. How many shells did Jamar find?

1 Understand Underline what you know. Circle what you need to find.

2 Plan How will I solve the problem?

3 Solve I will . . .

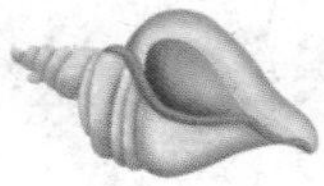

_______ shells

4 Check Is my answer reasonable? Explain.

Name ______________________

Processes & Practices

Apply the Strategy

Act it out to solve.

1. Lou picked 7 apples. Jordan picked 1 more apple than Lou. How many apples did Jordan pick?

_____ apples

2. Jan has 6 necklaces. Kim has the same number of necklaces. How many necklaces do they have altogether?

_____ necklaces

3. The clown sells toys in boxes of 2, 4, and 6. Nela's mom buys 2 boxes with 10 toys in all. Which two boxes does she buy?

boxes with _____ and _____ toys

Review the Strategies

Choose a strategy
- Act it out.
- Draw a diagram.
- Write a number sentence.

4. There are 5 boats in the lake. There are 6 boats out of the lake. How many boats are there in all?

_____ boats

5. The girls ride in 2 taxis. The boys ride in some taxis. There are 9 taxis in all. How many taxis do the boys ride in?

Part	Part
2	
Whole	
9	

_____ taxis

6. Alan counts 7 lamps. Kurt counts the same number of lamps. How many lamps do Kurt and Alan count?

_____ lamps

Name

My Homework

Lesson 6
Problem-Solving: Act It Out

Homework Helper

eHelp Need help? connectED.mcgraw-hill.com

There are 4 red pails. There are 3 more yellow pails than red pails. How many yellow pails are there?

1 Understand Underline what you know.
Circle what you need to find.

2 Plan How will I solve the problem?

3 Solve I will act it out.

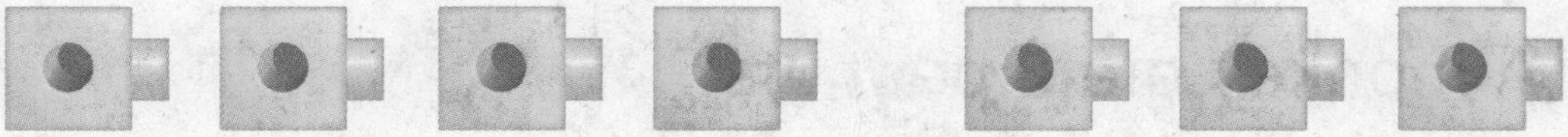

There are 7 yellow pails.

4 Check Is my answer reasonable?

Problem Solving

Underline what you know. Circle what you need to find. Act out the problem to solve. Use dry cereal.

1. A hot dog vendor sold 9 hot dogs on Monday. She sold the same number of hot dogs on Tuesday. How many hot dogs did she sell in all?

_____ hot dogs

2. There are 4 people jogging. There are 5 more people walking than jogging. How many people are walking?

_____ people

3. A monkey ate 5 peanuts. An elephant ate some peanuts. They ate 9 peanuts in all. How many peanuts did the elephant eat?

_____ peanuts

Math at Home Take advantage of problem-solving opportunities during daily routines such as riding in the car, doing laundry, putting away groceries, planning schedules, and so on.

Name ....

Make 10 to Add

Lesson 7

ESSENTIAL QUESTION
How do I use strategies to add numbers?

Explore and Explain

Watch Tools

Write your answer here. ______

Teacher Directions: Use counters to model. There are 9 counters in the yellow ten-frame. There are 5 counters in the purple ten-frame. Move counters to make 10. Color the boxes used. Write how many counters there are in all.

See and Show

You can make a 10 to help you add.

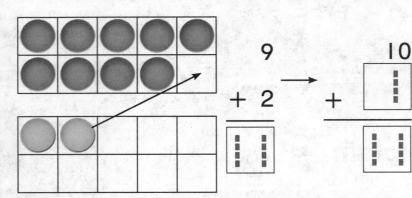

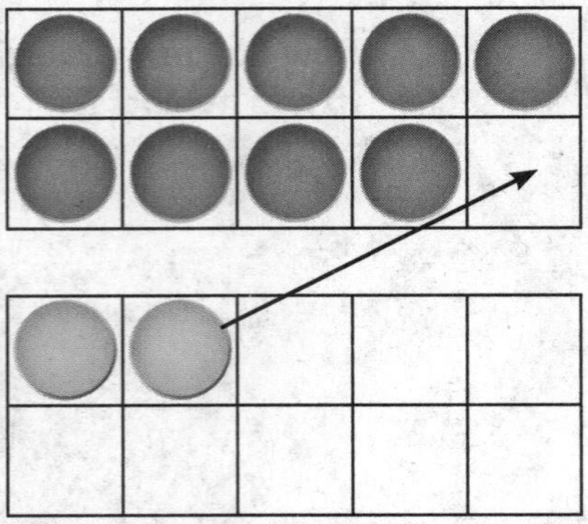

$$\begin{array}{r} 9 \\ +\ 2 \\ \hline 11 \end{array} \rightarrow \begin{array}{r} 10 \\ +\ 1 \\ \hline 11 \end{array}$$

Helpful Hint
Move 1 counter up to make a 10. 9 + 2 is the same as 10 + 1.

Use Work Mat 2 and counters. Make a ten to add.

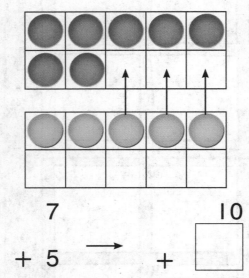

1.

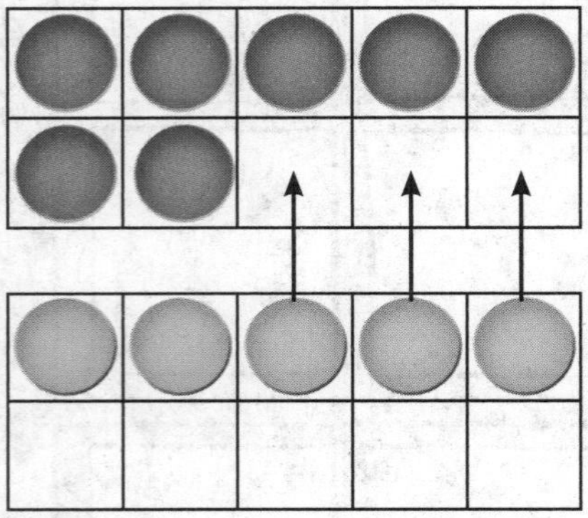

$$\begin{array}{r} 7 \\ +\ 5 \\ \hline \square \end{array} \rightarrow \begin{array}{r} 10 \\ +\ \square \\ \hline \square \end{array}$$

2.

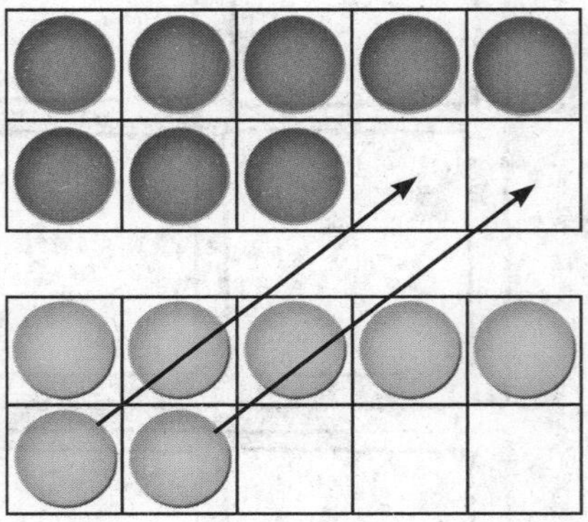

$$\begin{array}{r} 8 \\ +\ 7 \\ \hline \square \end{array} \rightarrow \begin{array}{r} 10 \\ +\ \square \\ \hline \square \end{array}$$

Talk Math Why is it helpful to make a 10 on a ten-frame when finding sums greater than 10?

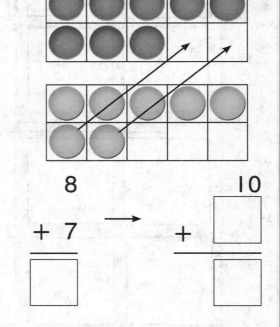

Name ....

On My Own

Use Work Mat 2 and [counters]. Make a ten to add.

3.

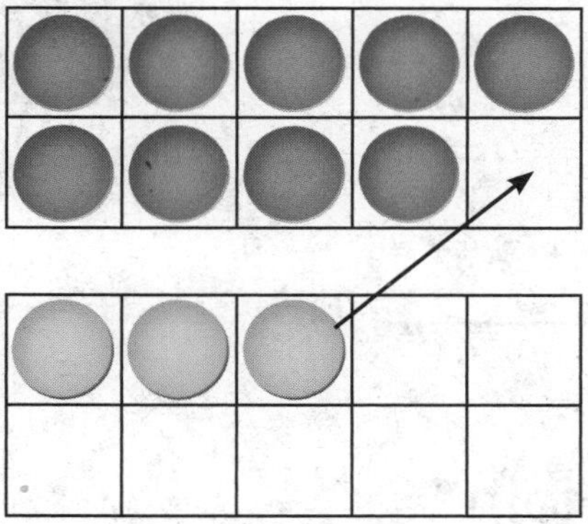

$9 + 3 = \square \rightarrow 10 + \square = \square$

4.

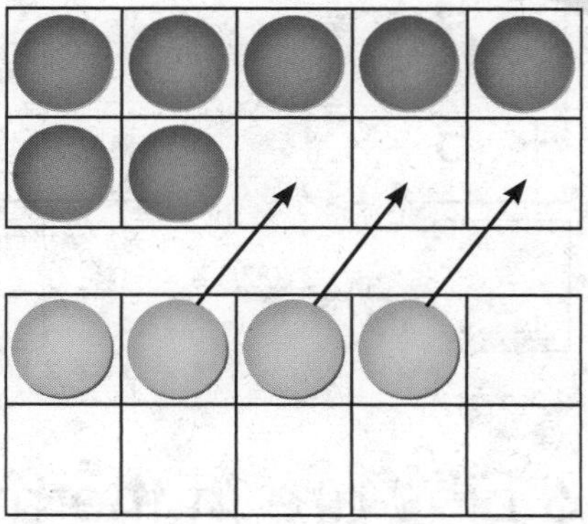

$7 + 4 = \square \rightarrow 10 + \square = \square$

5. $9 + 6 = \square \rightarrow 10 + \square = \square$

6. $8 + 5 = \square \rightarrow 10 + \square = \square$

7. $8 + 4 = \square \rightarrow 10 + \square = \square$

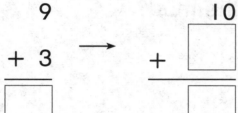

8. $7 + 6 = \square \rightarrow 10 + \square = \square$

Problem Solving

Processes & Practices

Use Work Mat 2 and ●● to solve.

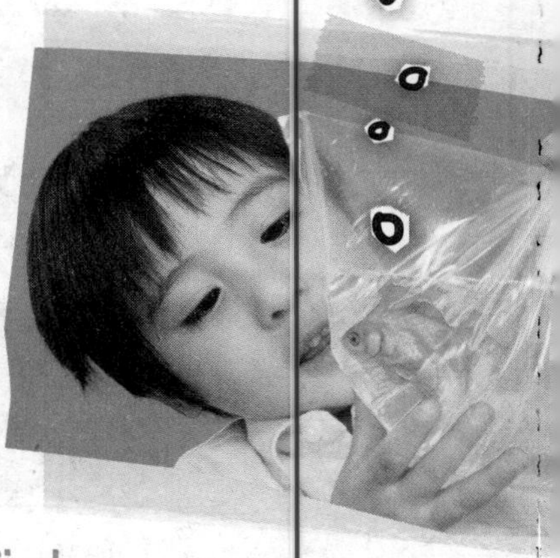

9. Don has 8 goldfish. He gets 5 more. How many goldfish does he have now?

$$\begin{array}{r} 8 \\ +\ 5 \\ \hline \square \end{array} \longrightarrow \begin{array}{r} 10 \\ +\ \square \\ \hline \square \end{array}$$

_____ goldfish

10. 6 pigs are in the mud. 5 more join them. How many pigs are in the mud?

$$\begin{array}{r} 6 \\ +\ 5 \\ \hline \square \end{array} \longrightarrow \begin{array}{r} 10 \\ +\ \square \\ \hline \square \end{array}$$

_____ pigs

Write Math Explain how to make a 10 to add using ten-frames.

Name ______________________

My Homework

Lesson 7
Make 10 to Add

Homework Helper

eHelp Need help? connectED.mcgraw-hill.com

You can make a 10 to add.

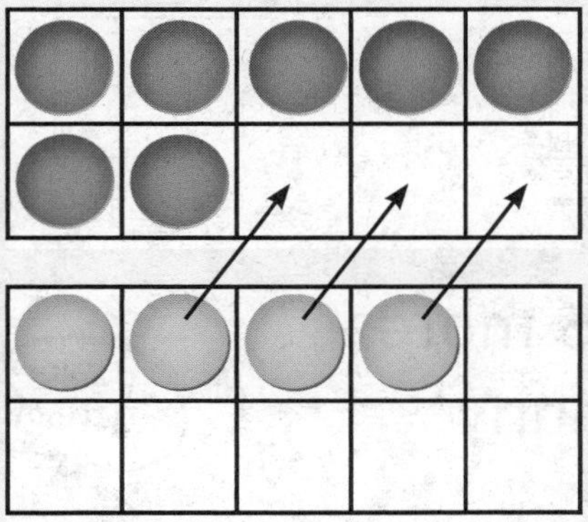

$$\begin{array}{r} 7 \\ +\ 4 \\ \hline 11 \end{array} \longrightarrow \begin{array}{r} 10 \\ +\ 1 \\ \hline 11 \end{array}$$

Helpful Hint
Move 3 counters up to make a 10.

Make a ten to add.

1.

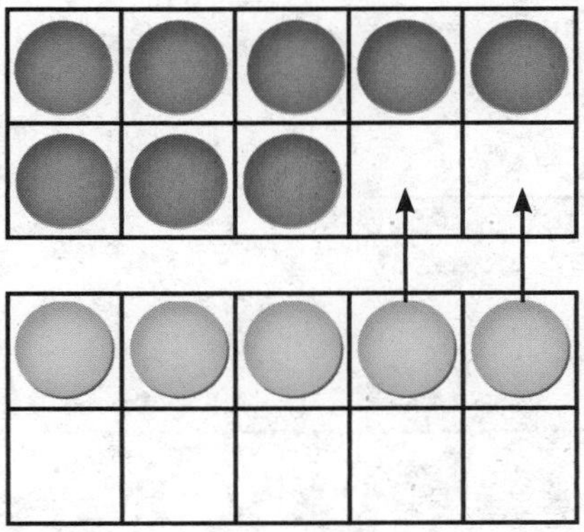

$$\begin{array}{r} 8 \\ +\ 5 \\ \hline \square \end{array} \longrightarrow \begin{array}{r} 10 \\ +\ \square \\ \hline \square \end{array}$$

2.

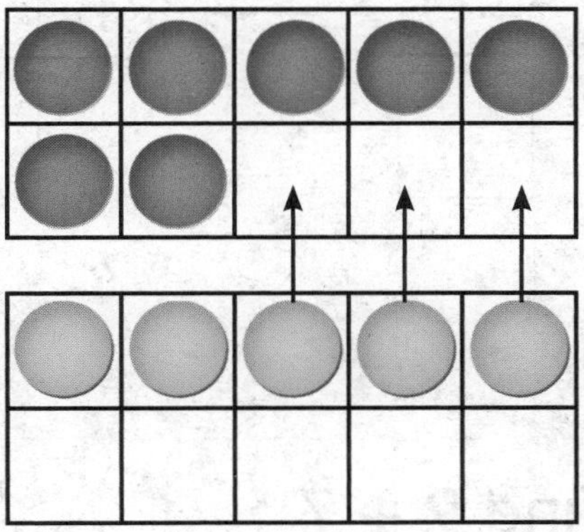

$$\begin{array}{r} 7 \\ +\ 5 \\ \hline \square \end{array} \longrightarrow \begin{array}{r} 10 \\ +\ \square \\ \hline \square \end{array}$$

Make a ten to add.

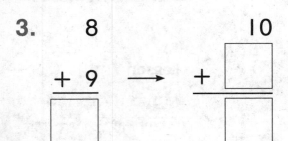

3. 8 + 9 = ☐ → 10 + ☐ = ☐

4. 9 + 4 = ☐ → 10 + ☐ = ☐

5. 7 + 8 = ☐ → 10 + ☐ = ☐

6. 8 + 6 = ☐ → 10 + ☐ = ☐

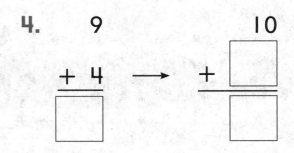

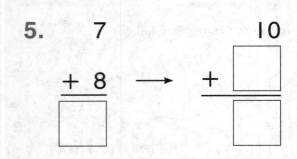

7. There are 7 squirrels in the park. 5 more squirrels come to the park. How many squirrels are in the park now?

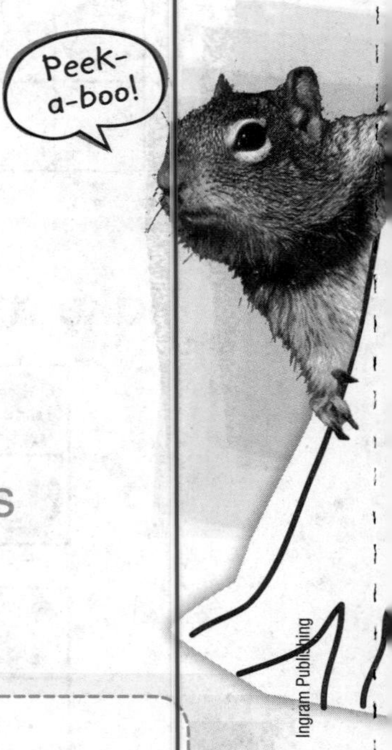

7 + 5 = ☐ → 10 + ☐ = ☐

_____ squirrels

Test Practice

8. Find 9 + 7.

○ 14 ○ 15 ○ 16 ○ 17

Math at Home Draw two ten-frames. Give your child addition problems with sums to 20 and pennies. Help him or her solve the problems using the ten-frames and pennies.

Name ....

Add in Any Order

Lesson 8

ESSENTIAL QUESTION
How do I use strategies to add numbers?

Explore and Explain

Watch Tools

4 + 3 = ______ 3 + 4 = ______

Teacher Directions: Use ●● to model. Show 4 + 3. Write the sum. Now change the order. Trace and color to match. Write the sum. Describe what you notice about the sums.

See and Show

Processes & Practices

You can change the order of the addends and get the same sum.

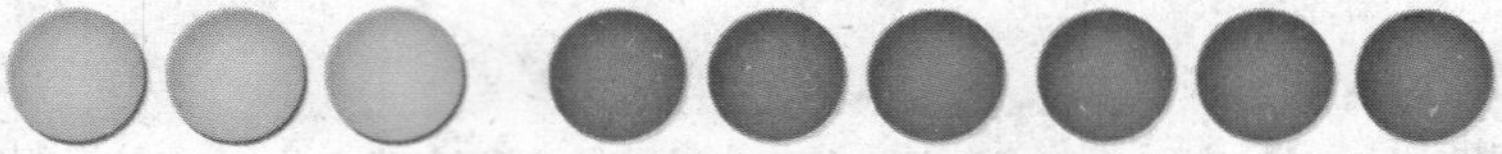

3 + 6 = 9

addend addend sum

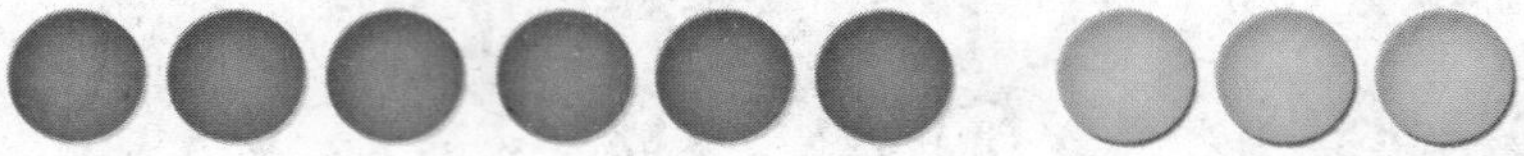

6 + 3 = 9

addend addend sum

Write the addends. Use  to add. Write the sum.

1. 

____ + ____ = ____

____ + ____ = ____

2. 

□
\+ □
―――
□

□
\+ □
―――
□

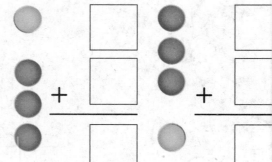

Talk Math Tell how you can show that 1 + 9 has the same sum as 9 + 1.

Name ....

Helpful Hint
If you change the order of the addends, you get the same sum.

On My Own

Write the addends. Add. Write the sum.

3. 

____ + ____ = ____

____ + ____ = ____

4. 

____ + ____ = ____

____ + ____ = ____

5. 

____ + ____ = ____

____ + ____ = ____

6. 

____ + ____ = ____

____ + ____ = ____

Add.

7. $\begin{array}{r} 6 \\ +\ 1 \\ \hline \end{array}$ $\quad$ $\begin{array}{r} 1 \\ +\ 7 \\ \hline \end{array}$ $\quad$ $\begin{array}{r} 1 \\ +\ 6 \\ \hline \end{array}$

8. $\begin{array}{r} 3 \\ +\ 3 \\ \hline \end{array}$ $\quad$ $\begin{array}{r} 5 \\ +\ 3 \\ \hline \end{array}$ $\quad$ $\begin{array}{r} 3 \\ +\ 5 \\ \hline \end{array}$

Problem Solving

Processes & Practices

Write two ways to add. Solve.

9. 4 ladybugs climb onto a leaf. 8 more join them. How many ladybugs are on the leaf?

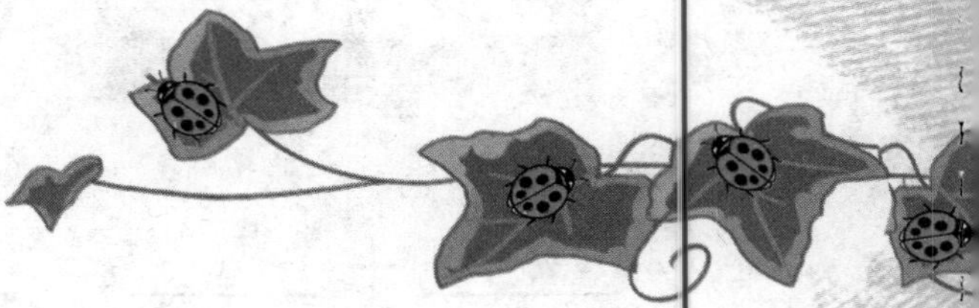

______ + ______

______ + ______ ______ ladybugs

10. 3 butterflies are in the garden. 0 butterflies join them. How many butterflies are in the garden?

______ + ______

______ + ______ ______ butterflies

Write Math Can you subtract in any order? Use ●●. Explain.

Name ....

My Homework

Lesson 8

Add in Any Order

Homework Helper

eHelp **Need help?** connectED.mcgraw-hill.com

You can change the order of the addends and get the same sum.

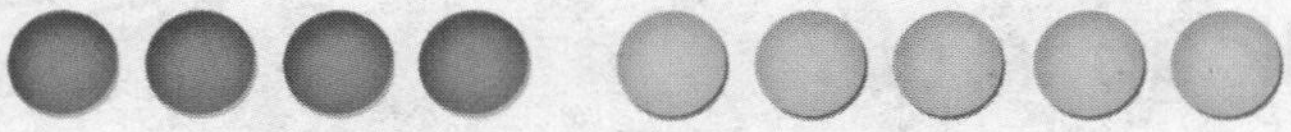

4 + 5 = 9

5 + 4 = 9

Practice

Write the addends. Then add.

1\.

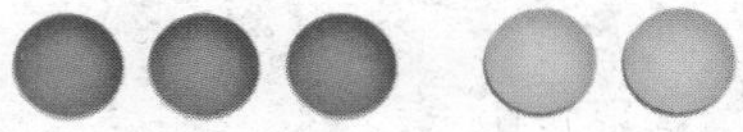

_____ + _____ = _____

_____ + _____ = _____

2\.

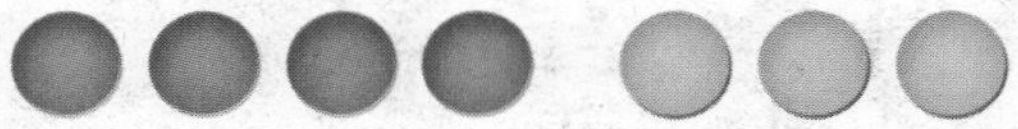

_____ + _____ = _____

_____ + _____ = _____

Add. Circle the addends that have the same sum.

3. 6 + 5 = ______ 5 + 6 = ______

4. 3 + 5 = ______ 3 + 6 = ______

Write two ways to add. Solve.

5. Greg walks up 7 steps to get to his apartment. Bill walks up 8 steps to get to his apartment. How many steps do they walk up in all?

______ + ______

______ + ______ ______ steps

Test Practice

6. Lynn walks 3 blocks to the library. Then she walks 8 blocks to her dance lesson. How many blocks does she walk in all?

9 ○ 10 ○ 11 ○ 12 ○

Math at Home Show your child 4 plates and 2 cups. Have him or her write two addition sentences about them.

Name ________________

Add Three Numbers

Lesson 9

ESSENTIAL QUESTION
How do I use strategies to add numbers?

Explore and Explain

Watch Tools

____ + ____ + ____ = ____

Teacher Directions: Use ●● to model. 6 pets live in building A. 4 pets live in building B. 4 pets live in building C. Add the number of pets in two of the buildings. Then add the number of pets in the third. Write the addition number sentence.

See and Show

Processes & Practices

You can group numbers and add in any order. Look for doubles. Look for numbers that make a ten.

Doubles	**Make a 10**
7 + (3) + (3) → 6	(7) + (3) + 3 → 10
Helpful Hint: 3 + 3 is a double.	Helpful Hint: 7 + 3 makes a 10.
7 + 6 = 13	10 + 3 = 13

Add the doubles or make a ten. Write that number. Add the other number to find the sum.

1. (6) + (4) + 3 = ____ □

2. (4) + 7 + (4) = ____ □

3.
```
  (3)
  (3)  □
+  4
----
```

4.
```
  (4)
   2   □
+ (4)
----
```

5.
```
  (4)
  (6)  □
+  2
----
```

Talk Math Tell how you would add the numbers 1 + 2 + 1.

Name ....

On My Own

Add the doubles or make a ten. Write that number. Add the other number to find the sum.

6. (2) + (2) + 3 = ____ ☐

7. 4 + (7) + (3) = ____ ☐

8. 3 + (9) + (1) = ____ ☐

9. (4) + (4) + 2 = ____ ☐

10. (2) + (8) + 1 = ____ ☐

11. 4 + (3) + (3) = ____ ☐

12.
(1)
(1) ☐
\+ 8
____

13.
(6)
(4) ☐
\+ 2
____

Problem Solving

Processes & Practices

Write an addition number sentence to solve.

14. On Monday, 8 ducks were in the pond. On Tuesday, 4 ducks were in the pond. On Wednesday, 4 ducks were in the pond. How many ducks were in the pond in all?

_____ + _____ + _____ = _____ ducks

15. An apartment building has 3 doors on the first floor, 5 doors on the second floor, and 5 doors on the third floor. How many doors in all?

_____ + _____ + _____ = _____ doors

Write Math Hayden says that $8 + 2 + 3 = 15$. Tell why Hayden is wrong. Make it right.

Name ....

My Homework

Lesson 9
Add Three Numbers

Homework Helper

eHelp

Need help? connectED.mcgraw-hill.com

You can group numbers and add in any order. Look for the doubles. Look for numbers that make a ten.

6 + (4) + (4)

[8]

4 + 4 is a double.
4 + 4 = 8.
Then add 6 + 8

6 + 8 = 14

(6) + (4) + 4

[10]

6 + 4 makes a 10.
Then add 10 + 4

10 + 4 = 14

Practice

Add the doubles or make a ten. Write that number. Add the other number to find the sum.

1. (5) + (5) + 2 = ______

2. (7) + 4 + (3) = ______

3. (2) + 9 + (2) = ______

4. (3) + (7) + 4 = ______

Write an addition number sentence to solve.

5. Bella sees 4 blue cars, 3 red cars, and 4 black cars. How many cars does she see in all?

_____ + _____ + _____ = _____ cars

6. Tom ate 1 slice of pizza. He also ate 9 carrots and 3 strawberries. How many pieces of food did he eat in all?

_____ + _____ + _____ = _____ pieces

7. Shannon buys 3 white shirts, 3 black shirts, and 4 green shirts. How many shirts does she buy in all?

_____ + _____ + _____ = _____ shirts

Test Practice

8. 2 + 8 + 4 = _____

10	11	12	14
○	○	○	○

Math at Home Place 3 crayons, 3 pencils, and 7 markers in front of your child. Have him or her identify how many there are of each and add the three numbers together to find how many there are in all. Have your child explain their work.

Name ....

Processes & Practices

Fluency Practice

Add.

1. 4 + 4 = ______
2. 2 + 1 = ______
3. 1 + 1 = ______
4. 1 + 8 = ______
5. 1 + 0 = ______
6. 4 + 2 = ______
7. 6 + 3 = ______
8. 2 + 7 = ______
9. 3 + 2 = ______
10. 2 + 5 = ______
11. 6 + 1 = ______
12. 5 + 4 = ______

13. $\begin{array}{r} 7 \\ +\ 3 \\ \hline \end{array}$

14. $\begin{array}{r} 3 \\ +\ 3 \\ \hline \end{array}$

15. $\begin{array}{r} 2 \\ +\ 2 \\ \hline \end{array}$

16. $\begin{array}{r} 6 \\ +\ 4 \\ \hline \end{array}$

17. $\begin{array}{r} 0 \\ +\ 9 \\ \hline \end{array}$

18. $\begin{array}{r} 0 \\ +\ 5 \\ \hline \end{array}$

19. $\begin{array}{r} 7 \\ +\ 1 \\ \hline \end{array}$

20. $\begin{array}{r} 1 \\ +\ 3 \\ \hline \end{array}$

21. $\begin{array}{r} 0 \\ +\ 0 \\ \hline \end{array}$

Fluency Practice

Add.

1. 0 + 8 = ______
2. 3 + 5 = ______
3. 7 + 9 = ______
4. 6 + 6 = ______
5. 1 + 6 = ______
6. 4 + 3 = ______
7. 6 + 7 = ______
8. 7 + 7 = ______
9. 3 + 8 = ______
10. 2 + 3 = ______
11. 9 + 0 = ______
12. 8 + 1 = ______

13. $\begin{array}{r} 4 \\ +\ 7 \\ \hline \end{array}$

14. $\begin{array}{r} 8 \\ +\ 4 \\ \hline \end{array}$

15. $\begin{array}{r} 10 \\ +\ 2 \\ \hline \end{array}$

16. $\begin{array}{r} 5 \\ +\ 2 \\ \hline \end{array}$

17. $\begin{array}{r} 9 \\ +\ 6 \\ \hline \end{array}$

18. $\begin{array}{r} 5 \\ +\ 6 \\ \hline \end{array}$

19. $\begin{array}{r} 10 \\ +\ 10 \\ \hline \end{array}$

20. $\begin{array}{r} 9 \\ +\ 9 \\ \hline \end{array}$

21. $\begin{array}{r} 9 \\ +\ 8 \\ \hline \end{array}$

Name ....

My Review

Chapter 3
Addition Strategies to 20

Vocabulary Check

Complete each sentence.

addends	**count on**	**doubles**
doubles minus 1	**doubles plus 1**	**number line**

1. Two addends that are the same number are called ________________.

2. You can use a number line to ________________ numbers to find the sum.

3. Any numbers being added together are called ________________.

4. To add with doubles plus one is called ________________.

Concept Check

Start with the greater number. Count on to add.

5. $7 + 2 =$ _____

6. $1 + 6 =$ _____

7. $4 + 3 =$ _____

8. $5 + 5 =$ _____

Make a ten to add.

9. 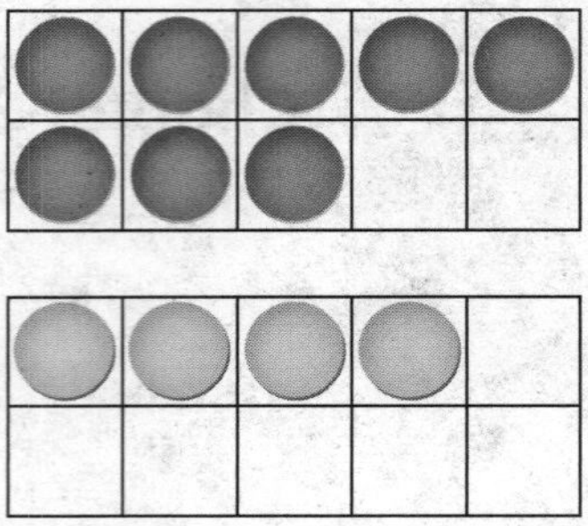

$$\begin{array}{r} 8 \\ +\ 4 \\ \hline \square \end{array} \longrightarrow \begin{array}{r} 10 \\ +\ \square \\ \hline \square \end{array}$$

10.

$$\begin{array}{r} 9 \\ +\ 9 \\ \hline \square \end{array} \longrightarrow \begin{array}{r} 10 \\ +\ \square \\ \hline \square \end{array}$$

11.

$$\begin{array}{r} 8 \\ +\ 3 \\ \hline \square \end{array} \longrightarrow \begin{array}{r} 10 \\ +\ \square \\ \hline \square \end{array}$$

Write the addends. Add.

12. 

_____ + _____ = _____

_____ + _____ = _____

13. 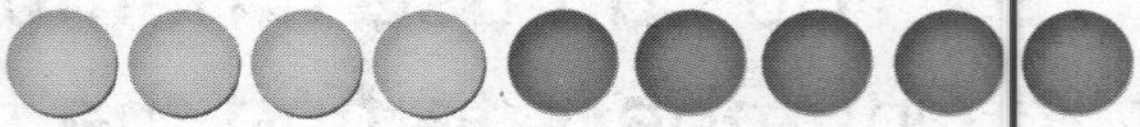

_____ + _____ = _____

_____ + _____ = _____

Add the doubles or make a ten. Write that number.
Add the other number to find the sum.

14. (3) + (3) + 5 = _____

☐

15. (1) + 7 + (9) = _____

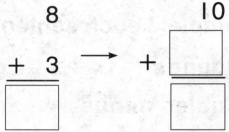

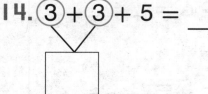

☐

Name

Problem Solving

16. Lea drank 2 glasses of milk. Juan drank 3 glasses of milk. How many glasses of milk did they drink in all?

_____ glasses

17. Jenny has 7 books from the library. Katy has 4 more books than Jenny. How many books does Katy have?

_____ books

Test Practice

18. 4 dogs play in the park. 4 dogs swim in the river. How many dogs are there in all?

4	6	8	10
○	○	○	○

Reflect

Chapter 3
Answering the Essential Question

Show the strategies you use to add.

ESSENTIAL QUESTION

How do I use strategies to add numbers?

Count On

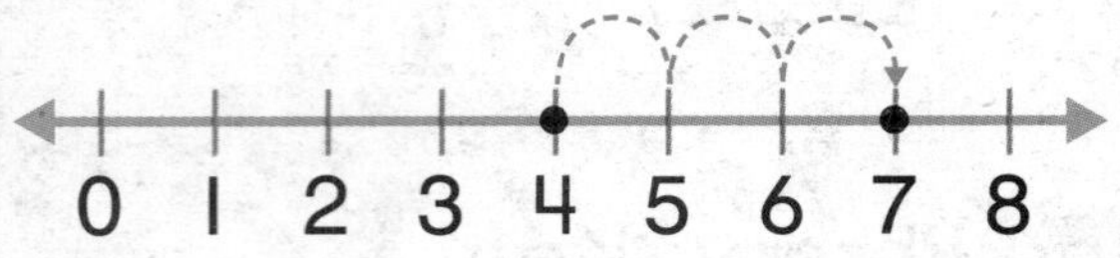

4 + 3 = ____

Doubles and Near Doubles

6 + 6 = ____

6 + 7 = ____

6 + 5 = ____

Add in Any Order

8 + 1 = ____

1 + 8 = ____

Add Three Numbers

(8) + 4 + (2) = ____

☐

Chapter

4 Subtraction Strategies to 20

ESSENTIAL QUESTION

What strategies can I use to subtract?

Let's Explore the Ocean!

Watch a video!

Name

Chapter 4 Project

My Story of the Chapter

1. Design a cover for your story of the chapter book below. Write a title for your book below.

2. Draw pictures, number sentences, or write vocabulary words that show the different things you learned in this chapter.

Name ....

Am I Ready?

1. Circle the minus sign.

+   −   =

2. Circle the equals sign.

+   −   =

Subtract.

3. $\begin{array}{r} 6 \\ -\ 0 \\ \hline \end{array}$

4. $\begin{array}{r} 5 \\ -\ 2 \\ \hline \end{array}$

5. $\begin{array}{r} 9 \\ -\ 8 \\ \hline \end{array}$

6. $\begin{array}{r} 4 \\ -\ 3 \\ \hline \end{array}$

7. $\begin{array}{r} 7 \\ -\ 3 \\ \hline \end{array}$

8. $\begin{array}{r} 3 \\ -\ 2 \\ \hline \end{array}$

9. Cross out 4 whales. Use the pictures to write a subtraction number sentence.

______ ◯ ______ ◯ ______

How Did I Do?

Shade the boxes to show the problems you answered correctly.

1	2	3	4	5	6	7	8	9

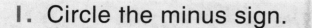

Name ....

My Math Words

Vocab

Review Vocabulary

false   related facts   true

Are the sets of number sentences related facts?
Circle true or false.

Number Sentences	True or False? Are they Related Facts?	
3 + 4 = 7 7 − 3 = 4	true	false
5 + 2 = 7 9 − 2 = 7	true	false
8 + 4 = 12 12 − 4 = 8	true	false

My Vocabulary Cards

Vocab abc

Processes & Practices

Lesson 4-1

count back

$6 - 2 = 4$

Lesson 4-7

fact family

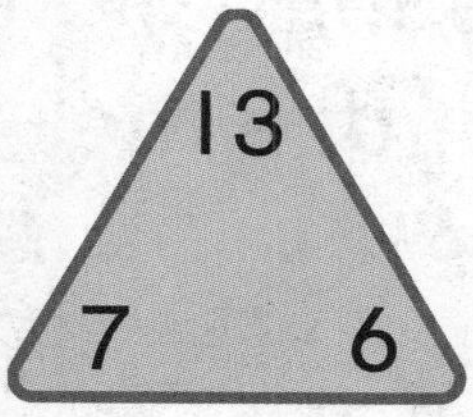

$6 + 7 = 13$

$7 + 6 = 13$

$13 - 6 = 7$

$13 - 7 = 6$

Lesson 4-8

missing addend

$5 + \square = 9$

$9 - 5 = \square$

Teacher Directions:
Ideas for Use

- Have students draw examples for each vocabulary card. Have them make drawings that are different from what is shown on the card.
- Ask students to use the blank cards to write their own vocabulary cards.

Addition and subtraction sentences that use the same numbers.

On a number line, start at the greater number and count back.

You can use subtraction facts to find a missing addend. The missing addend is 4.

My Foldable

FOLDABLES® Follow the steps on the back to make your Foldable.

$12 - 4 = 8$

$15 - 6 = 9$

$14 - 7 = 7$

$11 - 9 = 2$

$17 - 9 = 8$

FOLDABLES®
Study Organizer

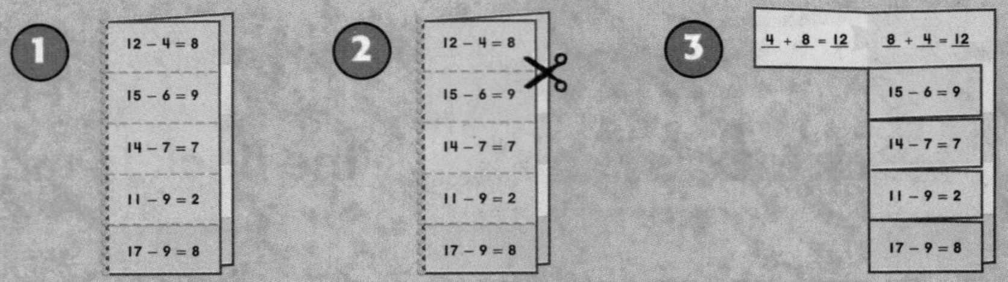

4 + 8 = 12	8 + 4 = 12
___ + ___ = ___	___ + ___ = ___
___ + ___ = ___	___ + ___ = ___
___ + ___ = ___	___ + ___ = ___
___ + ___ = ___	___ + ___ = ___

Name ______________________

Count Back 1, 2, or 3

Lesson 1

ESSENTIAL QUESTION
What strategies can I use to subtract?

Explore and Explain

Watch Tools

______ squirrels

Teacher Directions: Use [cube] to model. There are 5 squirrels playing. 3 of the squirrels run away. How many squirrels are still playing? Write the number.

See and Show

Processes & Practices

You can **count back** to subtract.

Start with 6. Count back 2.

6, 5, 4

6 − 2 = 4

Helpful Hint
Count back and take away cubes from the train. Write the numbers.

Count back to subtract. Use cubes to help.

1. Start with 8.

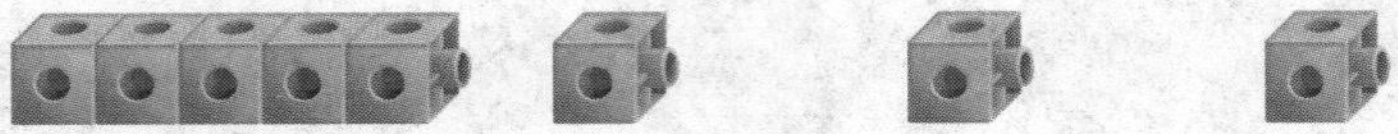

8, ____, ____, ____

8 − 3 = ____

2. 7, ____

7 − 1 = ____

3. 4, ____, ____

4 − 2 = ____

4. 12, ____, ____

12 − 2 = ____

5. 10, ____, ____, ____

10 − 3 = ____

Talk Math How do you count back to find 7 − 2?

Name ....

Your turn!

On My Own

Count back to subtract.
Use to help.

6. 11 − 2 = ______

7. 12 − 3 = ______

8. 3 − 2 = ______

9. 10 − 3 = ______

10. 9 − 3 = ______

11. 4 − 1 = ______

12. 6 − 1 = ______

13. 11 − 3 = ______

14. 7 − 3 = ______

15. 9 − 2 = ______

16. $\begin{array}{r} 10 \\ -\ 2 \\ \hline \end{array}$

17. $\begin{array}{r} 9 \\ -\ 2 \\ \hline \end{array}$

18. $\begin{array}{r} 8 \\ -\ 1 \\ \hline \end{array}$

19. $\begin{array}{r} 12 \\ -\ 2 \\ \hline \end{array}$

20. $\begin{array}{r} 10 \\ -\ 1 \\ \hline \end{array}$

21. $\begin{array}{r} 5 \\ -\ 3 \\ \hline \end{array}$

Problem Solving

Processes & Practices

Write a subtraction number sentence to solve.

22. There are 5 pelicans sitting on a rock. 1 of them flies away. How many pelicans are still sitting on the rock?

_____ – _____ = _____ pelicans

23. There are 11 boats at the dock. 3 of the boats leave. How many boats are still at the dock?

_____ – _____ = _____ boats

Write Math How do you count back to find 11 – 2? Explain.

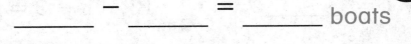

Name ....

My Homework

Lesson 1

Count Back 1, 2, or 3

Homework Helper

eHelp Need help? connectED.mcgraw-hill.com

Find 5 − 2. You can count back to subtract numbers.

Start with 5. Count back 2.

5, 4, 3

So, 5 − 2 = 3.

Helpful Hint
Count back and take away cubes from the train.

Practice

Count back to subtract.

1. 7, _____, _____

   7 − 2 = _____

2. 9, _____, _____, _____

   9 − 3 = _____

3. 12, _____, _____, _____

   12 − 3 = _____

4. 11, _____, _____

   11 − 2 = _____

Count back to subtract.

5. 11 − 3 = ______

6. 8 − 1 = ______

7. 5 − 2 = ______

8. 12 − 2 = ______

9. 10 − 3 = ______

10. 9 − 2 = ______

11. $\begin{array}{r} 8 \\ -\ 3 \\ \hline \end{array}$

12. $\begin{array}{r} 12 \\ -\ 1 \\ \hline \end{array}$

13. $\begin{array}{r} 11 \\ -\ 2 \\ \hline \end{array}$

14. Landon had 9 masks. He lost 3 of them. How many masks does he have left?

______ masks

Vocabulary Check

Circle the correct answer.

count back **count on**

15. Start at the number 6 and ________ by 2 to get the difference of 4.

Math at Home Write 12 − 3 = ___. Have your child write the answer and explain how to count back to solve the problem.

Name ....

Use a Number Line to Subtract

Lesson 2

ESSENTIAL QUESTION
What strategies can I use to subtract?

Explore and Explain

Watch Tools

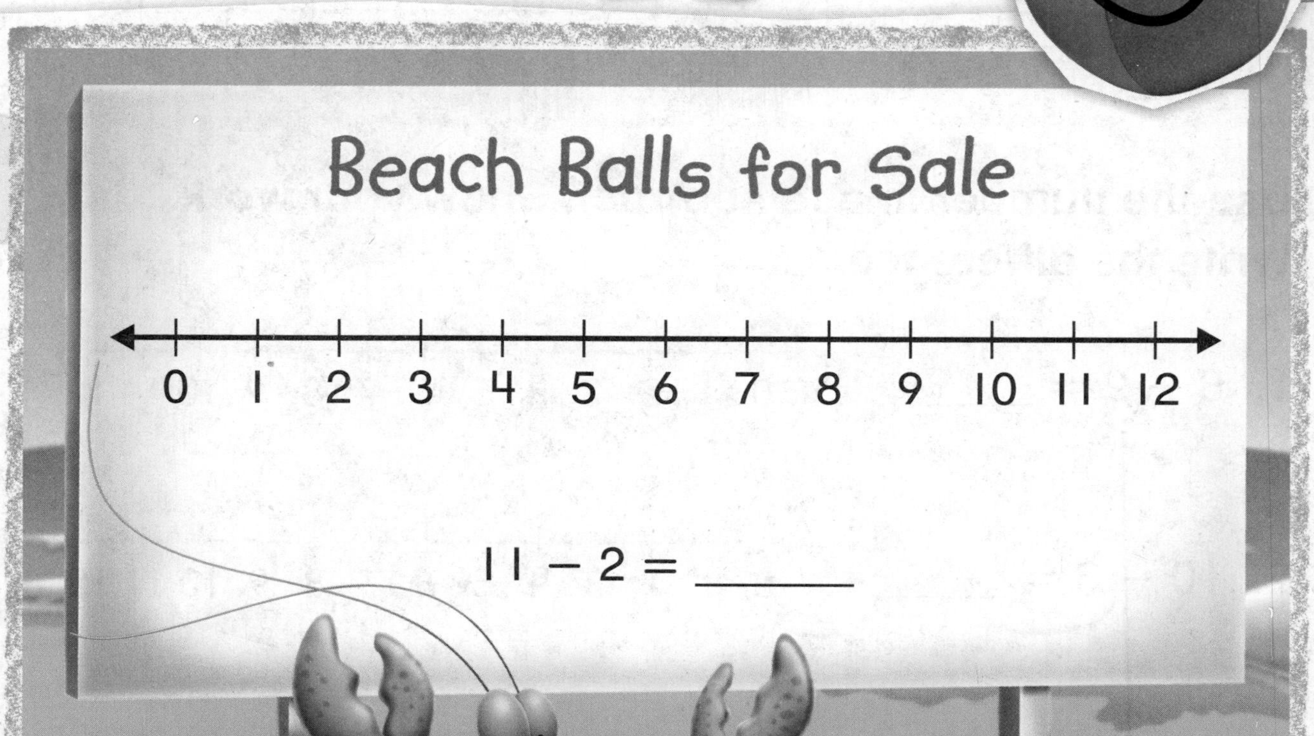

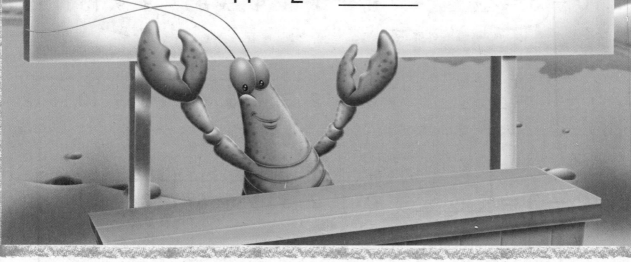

Teacher Directions: Use a paper clip to count back on the number line. A store has 11 beach balls to sell. It sells 2 of them. How many beach balls do they have left? Write the number.

See and Show

Processes & Practices

You can use a number line to subtract.

9 − 3 = <u>6</u>

Helpful Hint
Start at 9. Count back 3 to find the difference. 9, 8, 7, 6.

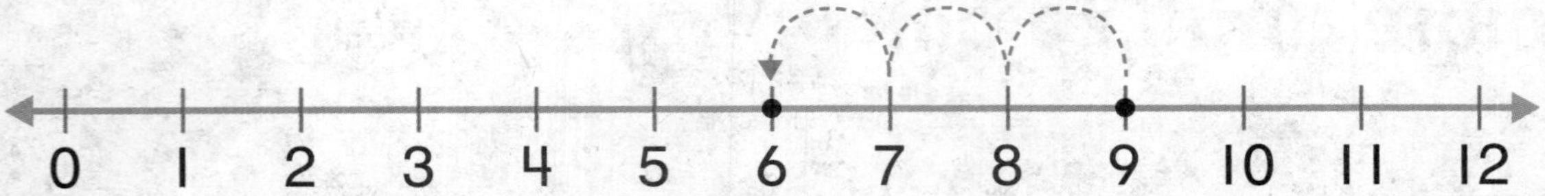

Use the number line to subtract. Show your work. Write the difference.

1. 8 − 2 = ______

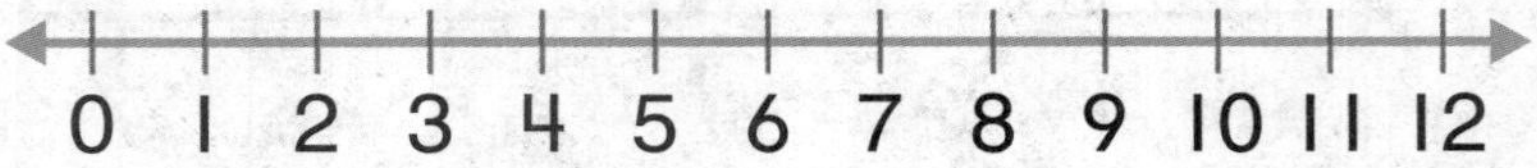

2. 10 − 3 = ______

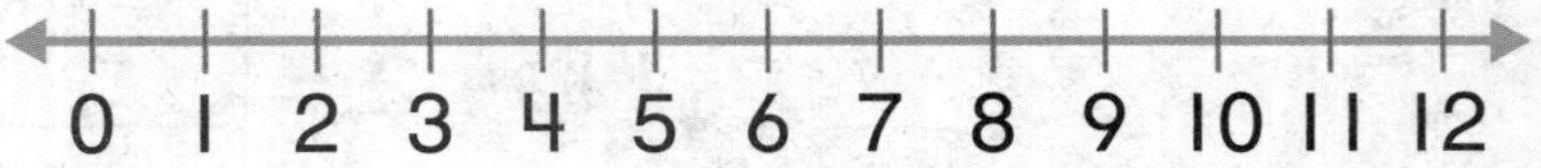

3. 5 − 1 = ______

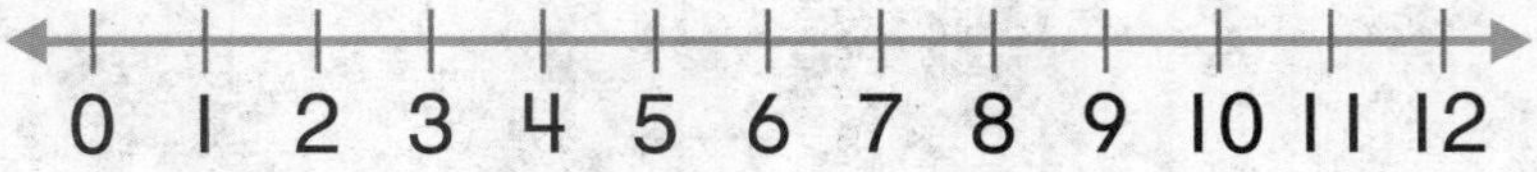

4. 12 − 3 = ______

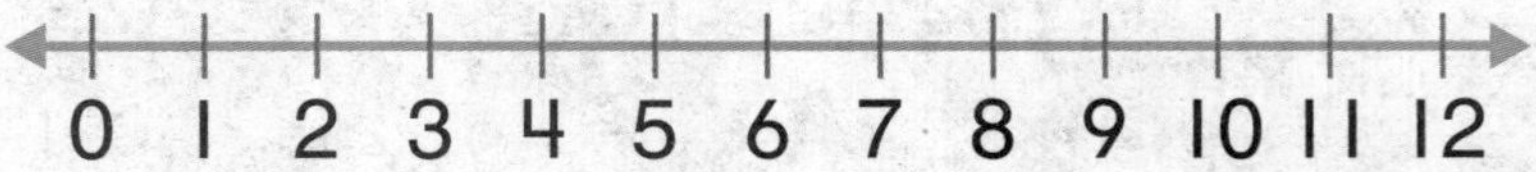

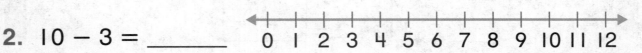

Talk Math Can you only use the number line to help you subtract numbers? Explain.

Name ....

On My Own

Use the number line to help you subtract. Write the difference.

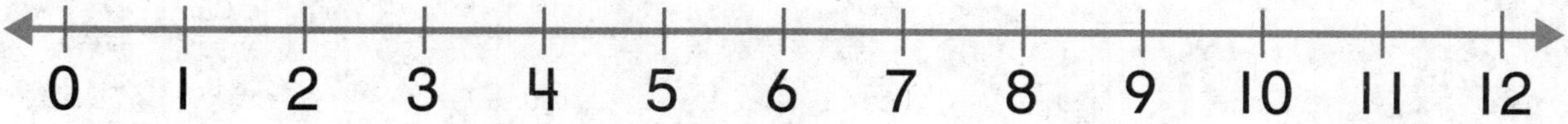

5. $\begin{array}{r} 8 \\ -\ 3 \\ \hline \end{array}$

6. $\begin{array}{r} 7 \\ -\ 3 \\ \hline \end{array}$

7. $\begin{array}{r} 10 \\ -\ 2 \\ \hline \end{array}$

8. $\begin{array}{r} 6 \\ -\ 2 \\ \hline \end{array}$

9. $\begin{array}{r} 7 \\ -\ 1 \\ \hline \end{array}$

10. $\begin{array}{r} 11 \\ -\ 3 \\ \hline \end{array}$

11. 12 − 1 = ______

12. 9 − 2 = ______

13. 10 − 1 = ______

14. 11 − 2 = ______

15. 5 − 3 = ______

16. 10 − 3 = ______

17. 12 − 2 = ______

18. 7 − 2 = ______

Problem Solving

Processes & Practices

19. Ava sees 12 sharks. 2 of them swim away. How many sharks does Ava see now?

______ sharks

20. There are 11 manatees near the shore. 3 of them swim away. How many manatees are still near the shore?

______ manatees

Write Math How do you use a number line to help you subtract? Explain.

Name ....

My Homework

Lesson 2
Use a Number Line to Subtract

Homework Helper

eHelp **Need help?** connectED.mcgraw-hill.com

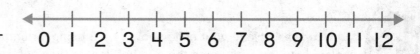

Use the number line to help you subtract.

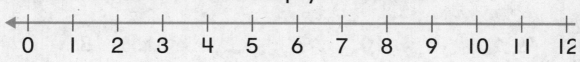

0 1 2 3 4 5 6 7 8 9 10 11 12

7 − 3 = 4

Practice

Use the number line to subtract. Show your work. Write the difference.

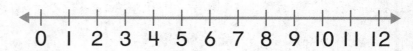

1. 10 − 3 = ______ 0 1 2 3 4 5 6 7 8 9 10 11 12

2. 6 − 2 = ______ 0 1 2 3 4 5 6 7 8 9 10 11 12

3. 12 − 3 = ______ 0 1 2 3 4 5 6 7 8 9 10 11 12

4. 5 − 2 = ______ 0 1 2 3 4 5 6 7 8 9 10 11 12

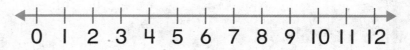

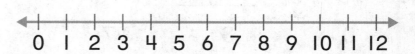

Use the number line to help you subtract.
Write the difference.

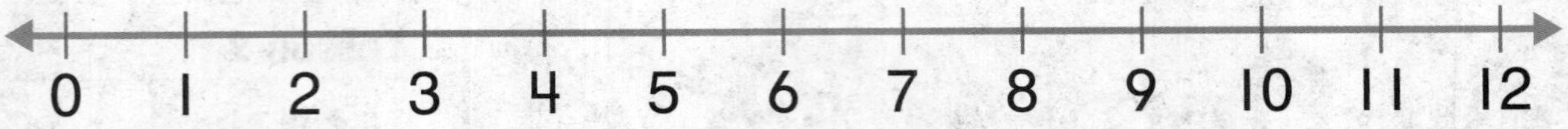

5. 11 − 2 = ______

6. 6 − 1 = ______

7. 9 − 3 = ______

8. 12 − 1 = ______

9. $\begin{array}{r} 6 \\ -\ 1 \\ \hline \end{array}$

10. $\begin{array}{r} 7 \\ -\ 3 \\ \hline \end{array}$

11. $\begin{array}{r} 8 \\ -\ 2 \\ \hline \end{array}$

12. There are 10 sharks swimming. 2 of the sharks swim away. How many sharks are still swimming?

______ sharks

Test Practice

13. 10 − 3 = ______

0 ○ 3 ○ 7 ○ 8 ○

Math at Home Have your child show 11 − 3 using a number line. Have him or her explain how they are using the number line as they subtract.

Name ______________________

Use Doubles to Subtract

Lesson 3

ESSENTIAL QUESTION
What strategies can I use to subtract?

All aboard!

Explore and Explain

Watch Tools

_____ − _____ = _____
Write your subtraction sentence here.

Teacher Directions: Use cubes to model. A train with 4 red train cars and 4 yellow train cars is on the tracks. The red train cars go to the right and the yellow train cars go to the left. Write the subtraction number sentence.

See and Show

Processes &Practices

You know how to use doubles facts to add.

4 + 4 = 8

Helpful Hint
Think 4 + 4 = 8,
so 8 − 4 = 4.

You can also use doubles facts to subtract.

8 − 4 = 4

Add the doubles facts. Then subtract.

1. 2 + 2 = ______ 4 − 2 = ______

2. 3 + 3 = ______ 6 − 3 = ______

3. $\begin{array}{r} 5 \\ +\ 5 \\ \hline \end{array}$ $\begin{array}{r} 10 \\ -\ 5 \\ \hline \end{array}$

4. $\begin{array}{r} 1 \\ +\ 1 \\ \hline \end{array}$ $\begin{array}{r} 2 \\ -\ 1 \\ \hline \end{array}$

Talk Math How can using doubles facts help you subtract?

Name ....

On My Own

Add the doubles facts. Then subtract.

5. 10 + 10 = ______

20 − 10 = ______

6. 4 + 4 = ______

8 − 4 = ______

7. 8 + 8 = ______ 16 − 8 = ______

8. 6 + 6 = ______ 12 − 6 = ______

9. 5 + 5 = ______ 10 − 5 = ______

10. 9 + 9 = ______ 18 − 9 = ______

Add or subtract. Draw lines to match the doubles facts.

11. 10 + 10 = ______ 14 − 7 = ______

12. 9 + 9 = ______ 16 − 8 = ______

13. 8 + 8 = ______ 18 − 9 = ______

14. 7 + 7 = ______ 20 − 10 = ______

Problem Solving

Processes & Practices

15. Bella sees 18 crabs on the beach. 9 of the crabs go in the ocean. How many crabs does Bella now see on the beach?

______ crabs

16. Jose found 14 shells on the beach. He gave 7 of the shells to his brother. How many shells does Jose have left?

Sally sells seashells by the seashore!

______ shells

HOT Problem Chase wrote $13 - 6 = 7$ on the board to show a doubles fact. Tell why Chase is wrong. Make it right.

Name ____________________

My Homework

Lesson 3

Use Doubles to Subtract

Homework Helper

eHelp **Need help?** connectED.mcgraw-hill.com

You can use doubles facts to help you subtract.

5 + 5 = 10

10 − 5 = 5

Practice

Add the doubles facts. Then subtract.

1. 4 + 4 = ______

8 − 4 = ______

2. 9 + 9 = ______

18 − 9 = ______

3. 8 + 8 = ______

16 − 8 = ______

4. 7 + 7 = ______

14 − 7 = ______

5. 3 + 3 = ______

6 − 3 = ______

6. 1 + 1 = ______

2 − 1 = ______

Add the doubles facts. Then subtract.

7. $\begin{array}{r} 5 \\ +\ 5 \\ \hline \end{array}$ $\begin{array}{r} 10 \\ -\ 5 \\ \hline \end{array}$

8. $\begin{array}{r} 2 \\ +\ 2 \\ \hline \end{array}$ $\begin{array}{r} 4 \\ -\ 2 \\ \hline \end{array}$

9. $\begin{array}{r} 6 \\ +\ 6 \\ \hline \end{array}$ $\begin{array}{r} 12 \\ -\ 6 \\ \hline \end{array}$

10. $\begin{array}{r} 10 \\ +10 \\ \hline \end{array}$ $\begin{array}{r} 20 \\ -10 \\ \hline \end{array}$

Subtract.

11. Camila has 6 pairs of sunglasses. She breaks 3 of them. How many pairs of sunglasses does she have left?

_____ sunglasses

Test Practice

12. Find the matching doubles fact.

9 + 9 = 18

16 − 9 = 7	18 − 9 = 9	7 + 7 = 14	18 − 9 = 8
○	○	○	○

Math at Home Write a doubles fact such as 7 + 7 = 14. Ask your child to give you a related subtraction fact.

Name ....

Real World

Problem Solving

STRATEGY: Write a Number Sentence

Lesson 4

ESSENTIAL QUESTION
What strategies can I use to subtract?

Watch

There are 6 seagulls flying over the ocean. 2 of the seagulls land in the ocean. How many seagulls are still flying?

1 Understand
Underline what you know. Circle what you need to find.

2 Plan
How will I solve the problem?

3 Solve
I will write a number sentence.

_____ ◯ _____ ◯ _____ seagulls

4 Check
Is my answer reasonable? Explain.

Practice the Strategy

There are 11 children building sand castles on the beach. 3 of the children go home. How many children are still building sand castles?

1 Understand Underline what you know.
Circle what you need to find.

2 Plan How will I solve the problem?

3 Solve I will...

_____ ◯ _____ ◯ _____ children

4 Check Is my answer reasonable? Explain.

I'm tired! That was hard work!

Name ....

Processes & Practices

Apply the Strategy

Write a subtraction number sentence to solve.

1. Isaac bought 12 shells. He lost 6 of the shells. How many shells does Isaac have now?

_____ ◯ _____ ◯ _____ shells

2. Juan found 11 starfish on the beach. He gave 4 of them to his sister. How many starfish does Juan have now?

_____ ◯ _____ ◯ _____ starfish

3. There are 10 sea horses swimming together. 8 of them swim away. How many sea horses are still swimming together?

_____ ◯ _____ ◯ _____ sea horses

Review the Strategies

Choose a strategy

- Write a number sentence.
- Draw a diagram.
- Act it out.

4. Brody saw 12 jellyfish in the ocean. Hunter saw 7 jellyfish there. How many more jellyfish did Brody see than Hunter?

______ jellyfish

5. There are 5 alligators swimming. 3 of them get out of the water. How many of the alligators are still swimming?

______ alligators

6. Lina had 10 sandals. She gave away 6 of them. How many sandals does Lina have left?

______ sandals

Name ____________________

My Homework

Lesson 4
Problem Solving: Write a Number Sentence

Homework Helper

eHelp **Need help?** connectED.mcgraw-hill.com

There are 12 dolphins swimming together. 6 of the dolphins swim away. How many dolphins are still swimming together?

1 Understand Underline what you know. Circle what you need to find.

2 Plan How will I solve the problem?

3 Solve I will write a number sentence.

$$12 - 6 = 6$$

6 dolphins are still swimming together.

4 Check Is my answer reasonable?

Problem Solving

Underline what you know. Circle what you need to find. Write a subtraction number sentence to solve.

1. There are 9 flamingos in the water. 5 flamingos get out. How many flamingos are still in the water?

_____ ◯ _____ ◯ _____ flamingos

2. 13 stingrays are swimming together. 4 of them swim away. How many stingrays are still swimming together?

_____ ◯ _____ ◯ _____ stingrays

3. Sydney saw 11 hermit crabs on the beach. Noah saw 6 hermit crabs there. How many more hermit crabs did Sydney see?

_____ ◯ _____ ◯ _____ hermit crabs

Math at Home Give your child a subtraction problem about things around the house. Have him or her write a subtraction number sentence to solve the problem.

Name ....

Check My Progress

Vocabulary Check

Vocab

Circle the correct answer.

count back **count on**

1. Start at a number and ________________ to subtract.

Concept Check

Count back to subtract.

2. $9 - 1 =$ ____

3. $11 - 3 =$ ____

4. $5 - 2 =$ ____

5. $7 - 3 =$ ____

6. $10 - 1 =$ ____

7. $8 - 3 =$ ____

8. $\begin{array}{r} 12 \\ -\ 2 \\ \hline \end{array}$

9. $\begin{array}{r} 6 \\ -\ 2 \\ \hline \end{array}$

10. $\begin{array}{r} 11 \\ -\ 2 \\ \hline \end{array}$

Use the number line to subtract. Write the difference.

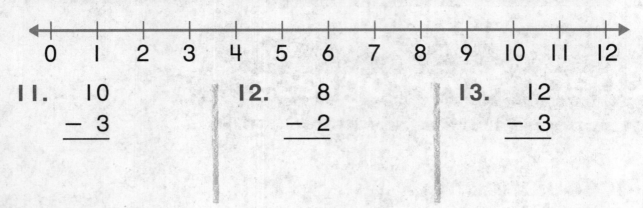

0 1 2 3 4 5 6 7 8 9 10 11 12

11. $\begin{array}{r} 10 \\ -\ 3 \\ \hline \end{array}$

12. $\begin{array}{r} 8 \\ -\ 2 \\ \hline \end{array}$

13. $\begin{array}{r} 12 \\ -\ 3 \\ \hline \end{array}$

Add the doubles facts. Then subtract.

14. $\begin{array}{r} 4 \\ +\ 4 \\ \hline \end{array}$ $\begin{array}{r} 8 \\ -\ 4 \\ \hline \end{array}$

15. $\begin{array}{r} 9 \\ +\ 9 \\ \hline \end{array}$ $\begin{array}{r} 18 \\ -\ 9 \\ \hline \end{array}$

Write a subtraction number sentence to solve.

16. There are 10 bikes parked at the beach. 2 children leave with their bikes. How many bikes are still at the beach?

____ − ____ = ____ bikes

Test Practice

17. There are 14 whales swimming together. 7 of them swim away. How many whales are still swimming together?

6 whales ○ 7 whales ○ 9 whales ○ 21 whales ○

Name ....

Make 10 to Subtract

Lesson 5

ESSENTIAL QUESTION
What strategies can I use to subtract?

Explore and Explain

Tools

13 − 7 =

3 | 4

13 − 3 = 10

10 − 4 = 6

Teacher Directions: Use [cube] to model. There are 13 coconuts on the island. 7 of the coconuts roll into the ocean. How many coconuts are left on the island? Trace the numbers. Draw the number of coconuts that are left on the island.

See and Show

Processes & Practices

To subtract, first take apart a number to make a 10. Then subtract.

16 − 9

6 | 3

16 − 6 = 10

10 − 3 = 7

So, 16 − 9 = 7.

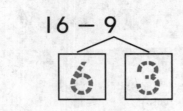

Helpful Hint
Think 16 − 6 = 10.
So, break apart 9 as 6 and 3.

16 − 6 = 10 and
10 − 3 = 7

Use  and . Take apart the number to make a 10. Then subtract.

1. 12 − 7

☐ ☐

12 − ____ = ☐

____ − ____ = ☐

So, 12 − 7 = ____.

2. 15 − 6

☐ ☐

15 − ____ = ☐

____ − ____ = ☐

So, 15 − 6 = ____.

Talk Math Explain how you can make a ten to find 13 − 7.

Name ____________________

On My Own

Use  and . Take apart the number to make a 10. Then subtract.

3. 17 − 9

☐ ☐

17 − ____ = ☐

____ − ____ = ☐

So, 17 − 9 = ____.

4. 12 − 8

☐ ☐

12 − ____ = ☐

____ − ____ = ☐

So, 12 − 8 = ____.

5. 11 − 5

☐ ☐

11 − ____ = ☐

____ − ____ = ☐

So, 11 − 5 = ____.

6. 14 − 7

☐ ☐

14 − ____ = ☐

____ − ____ = ☐

So, 14 − 7 = ____.

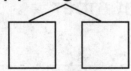

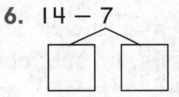

Problem Solving

Processes & Practices

7. Carly has 13 hats. She gives away 8 of them. How many hats does she have left?

_____ hats

8. There are 18 sea horses swimming in a group. 9 of the sea horses swim away. How many sea horses are still swimming in the group?

_____ sea horses

Write Math How do you take apart a number to subtract? Explain.

Name ______________________

My Homework

Lesson 5
Make 10 to Subtract

Homework Helper

eHelp Need help? connectED.mcgraw-hill.com

You can make a 10 to subtract more easily.

14 − 8

4 4

14 − 4 = 10

10 − 4 = 6

So, 14 − 8 = 6.

Helpful Hint
Think 14 − 4 = 10.
So, break apart
8 as 4 and 4.

14 − 4 = 10 and
10 − 4 = 6

Practice

Take apart the number to make a 10. Then subtract.

1. 17 − 9

[ ] [ ]

17 − ______ = [ ]

______ − ______ = [ ]

So, 17 − 9 = ______.

2. 13 − 7

[ ] [ ]

13 − ______ = [ ]

______ − ______ = [ ]

So, 13 − 7 = ______.

Take apart the number to make a 10. Then subtract.

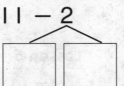

3. 11 − 2

11 − ______ = ☐

______ − ______ = ☐

So, 11 − 2 = ______.

4. 16 − 8

16 − ______ = ☐

______ − ______ = ☐

So, 16 − 8 = ______.

5. There are 18 coconuts on a tree. 9 coconuts fall off the tree. How many coconuts are still on the tree?

______ coconuts

Test Practice

6. Sadie collects 15 shells at the beach. She gives her brother 6 of them. How many shells does she have left?

18 shells ○ 12 shells ○ 11 shells ○ 9 shells ○

Math at Home Write 13 − 8 = ___ on a piece of paper. Have your child use buttons or marbles to help him or her subtract.

Name ....

Use Related Facts to Add and Subtract

Lesson 6

ESSENTIAL QUESTION
What strategies can I use to subtract?

Explore and Explain

Watch Tools

☐ + ☐ = ☐   ☐ − ☐ = ☐

Teacher Directions: Use [cube] to model. There are 6 players on one side of the net and 8 players on the other side. How many people are playing volleyball in all? Write the addition number sentence. Then write a related subtraction fact.

See and Show

Processes &Practices

Related facts use the same numbers.
These facts can help you add and subtract.
Find 11 − 5.

6 + 5 = 11

Helpful Hint
Use 6 + 5 = 11 to find 11 − 5 = 6.

11 − 5 = 6

Use related facts to add and subtract.

1. 7 + 9 = ____

16 − 7 = ____

2. 5 + 8 = ____

13 − 5 = ____

3.

$$\begin{array}{r} 5 \\ +\ 7 \\ \hline \end{array} \qquad \begin{array}{r} 12 \\ -\ 5 \\ \hline \end{array}$$

4.

$$\begin{array}{r} 8 \\ +\ 7 \\ \hline \end{array} \qquad \begin{array}{r} 15 \\ -\ 7 \\ \hline \end{array}$$

Talk Math Are the facts 1 + 5 = 6 and 6 − 1 = 5 related facts? How do you know?

Name ....

On My Own

Use related facts to add and subtract.

5. 9 + 6 = ____

15 − 9 = ____

6. 6 + 7 = ____

13 − 6 = ____

7. 3 + 9 = ____

12 − 3 = ____

8. 8 + 9 = ____

17 − 8 = ____

9. $\begin{array}{r} 6 \\ +\ 8 \\ \hline \end{array}$ $\begin{array}{r} 14 \\ -\ 6 \\ \hline \end{array}$

10. $\begin{array}{r} 8 \\ +\ 4 \\ \hline \end{array}$ $\begin{array}{r} 12 \\ -\ 4 \\ \hline \end{array}$

Subtract. Write an addition fact to check your subtraction.

11. 16 − 9 = ____

____ + ____ = ____

12. 12 − 7 = ____

____ + ____ = ____

13. 14 − 9 = ____

____ + ____ = ____

14. 11 − 4 = ____

____ + ____ = ____

Problem Solving

Processes & Practices

Write a subtraction number sentence. Then write a related addition fact.

15. Bailey sees 15 birds sitting on a rock. 7 of the birds fly away. How many of the birds are still on the rock?

_____ − _____ = _____ _____ + _____ = _____

16. Andre collects 10 shells. He loses 6 of them. How many shells does Andre have left?

_____ − _____ = _____ _____ + _____ = _____

Write Math How can related facts help you add and subtract? Explain.

Name ______________________________

My Homework

Lesson 6

Use Related Facts to Add and Subtract

Homework Helper

eHelp **Need help?** connectED.mcgraw-hill.com

Related facts can help you add and subtract.

$8 + 4 = 12$ $9 + 6 = 15$

$12 - 4 = 8$ $15 - 9 = 6$

Practice

Use related facts to add and subtract.

1. $5 + 9 =$ ______

$14 - 5 =$ ______

2. $9 + 5 =$ ______

$14 - 9 =$ ______

3. $6 + 5 =$ ______

$11 - 6 =$ ______

4. $9 + 9 =$ ______

$18 - 9 =$ ______

5. $\begin{array}{r} 8 \\ +\ 8 \\ \hline \end{array}$ $\begin{array}{r} 16 \\ -\ 8 \\ \hline \end{array}$

6. $\begin{array}{r} 8 \\ +\ 5 \\ \hline \end{array}$ $\begin{array}{r} 13 \\ -\ 8 \\ \hline \end{array}$

Subtract. Write an addition fact to check your subtraction.

7. 15 − 8 = ______

______ + ______ = ______

8. 17 − 9 = ______

______ + ______ = ______

Write a subtraction number sentence. Then write a related addition fact.

9. There are 16 lobsters swimming together. 7 of them swim away. How many lobsters are still swimming together?

______ − ______ = ______

______ + ______ = ______

Test Practice

10. Mark the related addition fact.

7 − 3 = 4

7 + 3 = 10	7 − 4 = 3	7 + 1 = 8	3 + 4 = 7
○	○	○	○

Math at Home Write an addition fact such as 3 + 9 = 12. Ask your child to write a related subtraction fact.

Name ____________________

Fact Families

Lesson 7

ESSENTIAL QUESTION
What strategies can I use to subtract?

Explore and Explain

Watch Tools

☐ + ☐ = ☐ ☐ − ☐ = ☐

☐ + ☐ = ☐ ☐ − ☐ = ☐

Teacher Directions: Use [cube] to model. There are 7 fish swimming together in a group. 3 more fish join them. Draw a picture to show the story. Write the missing numbers to show the fact family.

See and Show

Processes & Practices

Helpful Hint
5, 6, and 11 are a fact family.

Related facts make a **fact family.**

11
5 6

$5 + 6 = 11$ $11 - 5 = 6$

$6 + 5 = 11$ $11 - 6 = 5$

Add and subtract. Complete the fact family.

1.
9
2 7

$7 + 2 = \square$ $9 - 7 = \square$

$2 + 7 = \square$ $9 - 2 = \square$

2.
13
5 8

$5 + 8 = \square$ $13 - 5 = \square$

$8 + 5 = \square$ $13 - 8 = \square$

3.
14
6 8

$6 + 8 = \square$ $14 - 6 = \square$

$8 + 6 = \square$ $14 - 8 = \square$

Talk Math What fact family can you make with the numbers 4, 9, and 13?

Name ....

On My Own

In a fact family, all facts have the same numbers.

Add and subtract.
Complete the fact family.

4.

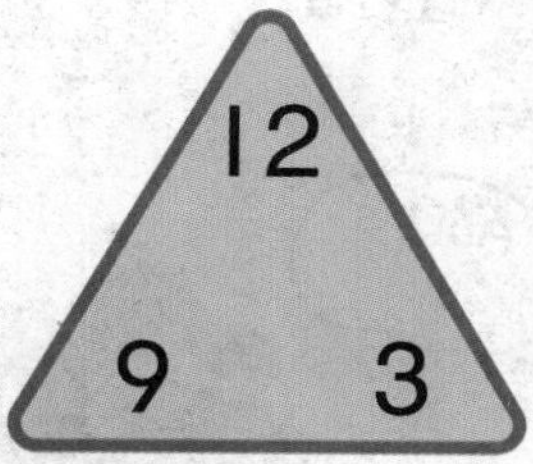

9 + 3 = ☐   12 − 9 = ☐

3 + 9 = ☐   12 − 3 = ☐

5.

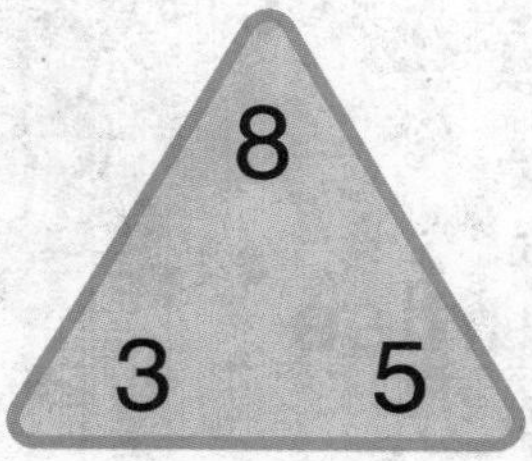

3 + 5 = ☐   8 − 3 = ☐

5 + 3 = ☐   8 − 5 = ☐

6.

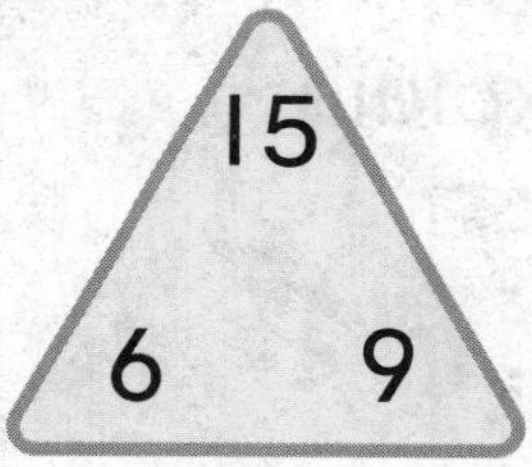

6 + 9 = ☐   15 − 9 = ☐

9 + 6 = ☐   15 − 6 = ☐

7.

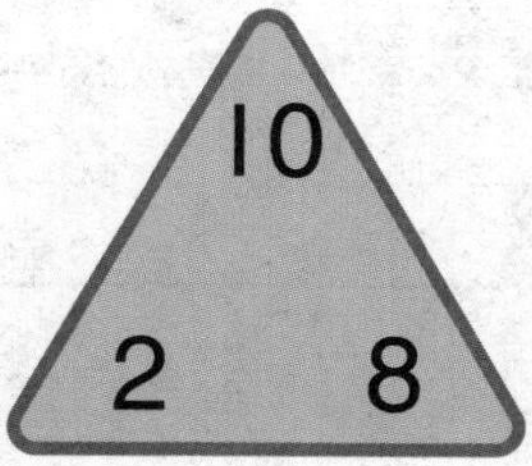

2 + 8 = ☐   10 − 2 = ☐

8 + 2 = ☐   10 − 8 = ☐

Problem Solving

Processes & Practices

8. Joe finds 12 starfish at the beach.
He gives 5 to his grandmother.
How many starfish does he have left?

______ starfish

9. There are 16 turtles in the ocean.
7 of those turtles get out. How many turtles are still in the ocean?

______ turtles

HOT Problem When I am subtracted from 17, the difference is 9. What number am I? Explain.

$$17 - \square = 9$$

Name ____________________

My Homework

Lesson 7
Fact Families

Homework Helper

eHelp **Need help?** connectED.mcgraw-hill.com

Related facts make up a fact family.

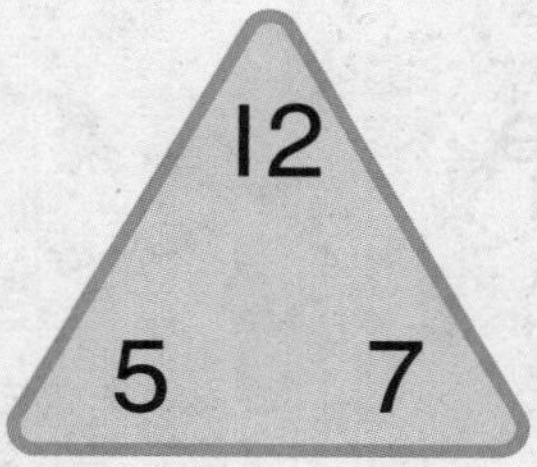

5 + 7 = 12

7 + 5 = 12

12 − 5 = 7

12 − 7 = 5

Practice

Add and subtract. Complete the fact family.

1.

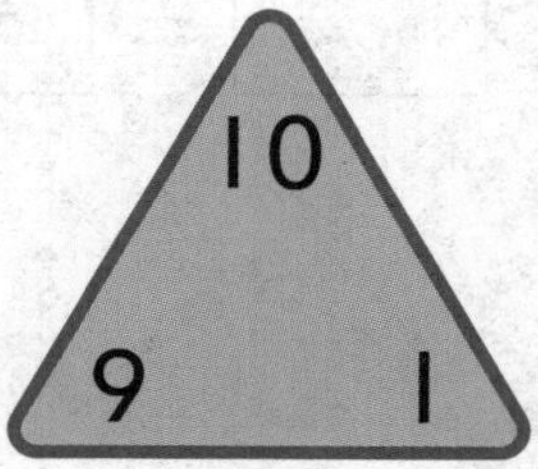

1 + 9 = ☐

9 + 1 = ☐

10 − 9 = ☐

10 − 1 = ☐

2.

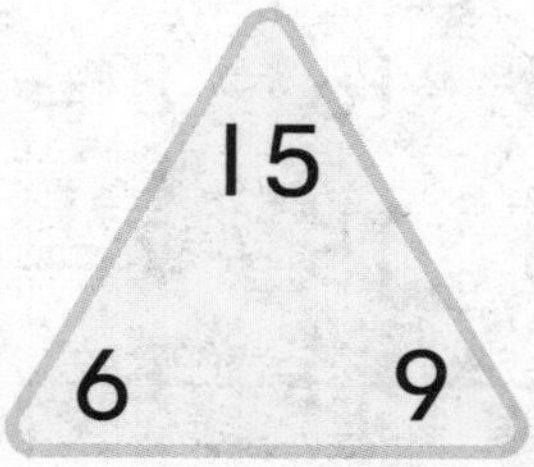

6 + 9 = ☐

9 + 6 = ☐

15 − 9 = ☐

15 − 6 = ☐

Complete the fact family.

3. Pedro sees 5 flamingos in the water. He sees 4 flamingos on the grass. How many flamingos did he see in all?

$\square + \square = \square$ $\square - \square = \square$

$\square + \square = \square$ $\square - \square = \square$

4. Hailey saw 9 crabs on the beach in the morning. She saw 8 more crabs in the afternoon. How many crabs did Hailey see in all?

$\square + \square = \square$ $\square - \square = \square$

$\square + \square = \square$ $\square - \square = \square$

Vocabulary Check

Vocab

5. Circle the **fact family.**

$2 + 3 = 5$	$4 - 2 = 2$	$7 + 8 = 15$	$15 - 8 = 7$
$3 + 1 = 4$	$5 - 1 = 4$	$8 + 7 = 15$	$15 - 7 = 8$

Math at Home Challenge your child to write all of the fact families that make 15.

Name ....

Missing Addends

Lesson 8

ESSENTIAL QUESTION
What strategies can I use to subtract?

Explore and Explain

Watch Tools

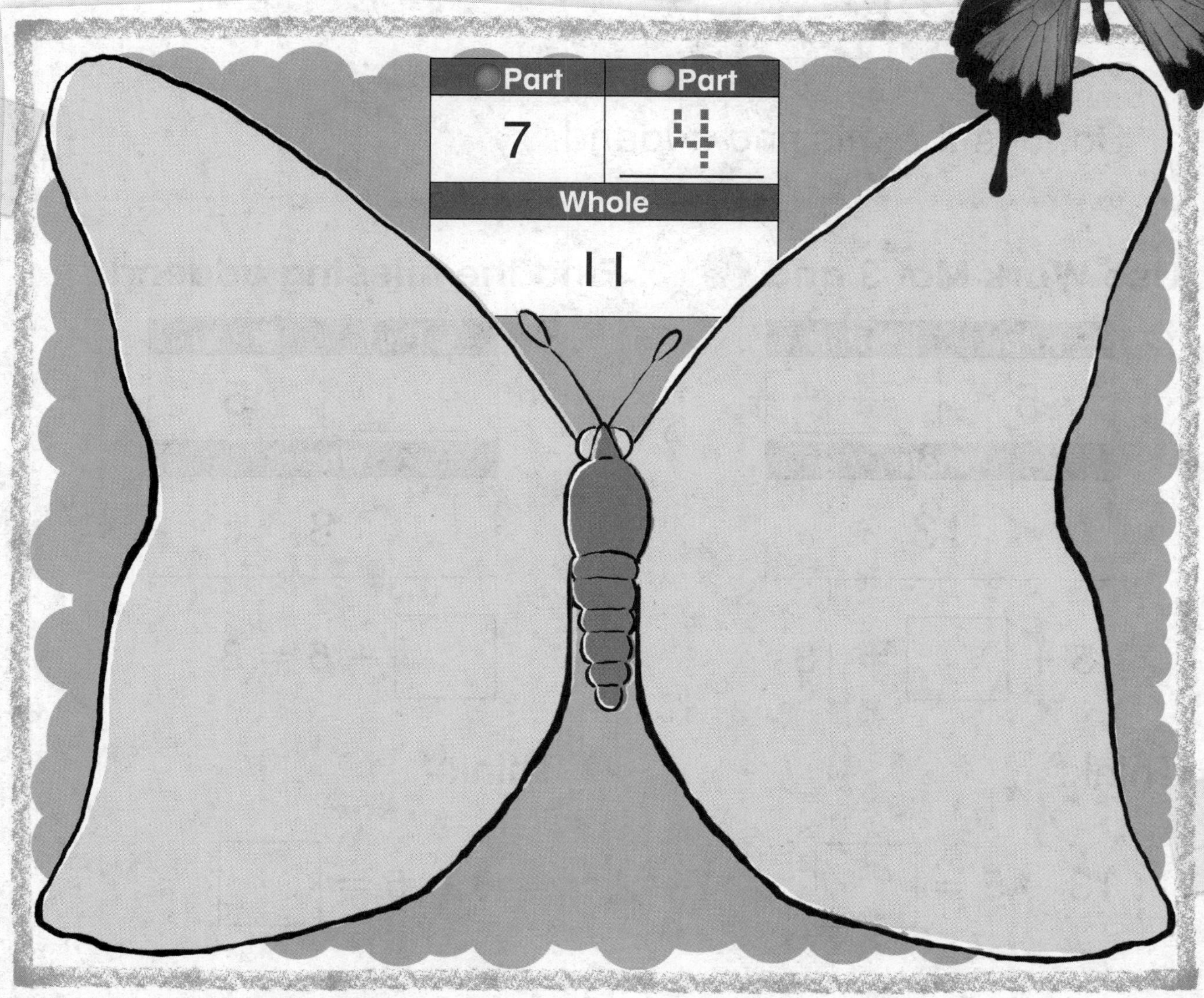

Part	Part
7	4
Whole	
11	

Teacher Directions: Use [counters] to model. The butterfly has 7 spots on the left wing and some more spots on its right wing. It has 11 spots in all. How many spots are on the right wing? Trace your counters to show the spots. Trace the missing part.

See and Show

Processes & Practices

Helpful Hint
To find the missing part, subtract the known part from the whole.

Use related facts to help you find a **missing addend**.

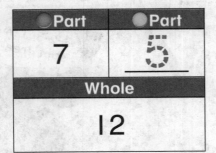

Part	Part
7	5
Whole	
12	

$7 + \boxed{5} = 12$

$12 - 7 = \boxed{5}$

So, 5 is the missing addend.

Use Work Mat 3 and ●●. Find the missing addend.

1.

Part	Part
5	___
Whole	
13	

$5 + \square = 13$

Think:

$13 - 5 = \square$

2.

Part	Part
___	5
Whole	
8	

$\square + 5 = 8$

Think:

$8 - 5 = \square$

Talk Math Explain how to find the missing addend in $\square + 5 = 14$.

Name

You can find missing addends by subtracting.

On My Own

Use Work Mat 3 and ●●. Find the missing addend.

3.

Part	Part
4	___
Whole	
10	

$4 + \square = 10$

$10 - 4 = \square$

4.

Part	Part
___	6
Whole	
14	

$\square + 6 = 14$

$14 - 6 = \square$

5. $8 + \square = 9$

$9 - 8 = \square$

6. $\square + 3 = 11$

$11 - 3 = \square$

7. $\square + 7 = 13$

$13 - 7 = \square$

8. $6 + \square = 15$

$15 - 6 = \square$

9. $9 + \square = 17$

$17 - 9 = \square$

10. $9 + \square = 16$

$16 - 9 = \square$

Problem Solving

Processes & Practices

11. Max has 5 shovels and some sand pails at the beach. He has 14 shovels and sand pails in all. How many sand pails does he have?

_____ sand pails

12. When I am added to 7, the sum is 12. What number am I?

7 12

_____

Write Math How do you use subtraction to find a missing addend in an addition problem? Explain.

Name ....

My Homework

Lesson 8
Missing Addends

Homework Helper

eHelp Need help? connectED.mcgraw-hill.com

You can use related facts to help you find a missing addend.

Part	Part
5	4
Whole	
9	

5 + [4] = 9

9 − 5 = [4]

Helpful Hint
To find the missing part, subtract the known part from the whole.

Practice

Find the missing addend.

1.

Part	Part
8	____
Whole	
11	

8 + [ ] = 11

11 − 8 = [ ]

2.

Part	Part
____	8
Whole	
16	

[ ] + 8 = 16

16 − 8 = [ ]

Find the missing addend.

3. $9 + \square = 18$

   $18 - 9 = \square$

4. $\square + 6 = 14$

   $14 - 6 = \square$

5. $\square + 8 = 15$

   $15 - 8 = \square$

6. $5 + \square = 11$

   $11 - 5 = \square$

7. There are 16 children flying kites on the beach. Some of the children go home. 9 children are still flying kites. How many children went home?

_____ children

Vocabulary Check

Vocab

8. Circle the number sentence that shows a **missing addend**.

   $4 + 8 = 12$ $\quad$ $7 + \square = 15$

Math at Home Ask your child to tell you the subtraction fact that will help him or her find the missing addend in 8 + □ = 15.

Name ______________________

Processes & Practices

Fluency Practice

Subtract.

1. 5 − 3 = ______
2. 18 − 9 = ______
3. 4 − 2 = ______
4. 10 − 6 = ______
5. 11 − 5 = ______
6. 7 − 0 = ______
7. 9 − 9 = ______
8. 4 − 1 = ______
9. 14 − 6 = ______
10. 3 − 0 = ______
11. 1 − 1 = ______
12. 8 − 3 = ______
13. 13 − 6 = ______
14. 6 − 3 = ______
15. 5 − 4 = ______
16. 16 − 8 = ______
17. 4 − 4 = ______
18. 15 − 6 = ______
19. 10 − 8 = ______
20. 17 − 8 = ______
21. 9 − 1 = ______
22. 7 − 3 = ______
23. 14 − 7 = ______
24. 10 − 5 = ______

Fluency Practice

Subtract.

1. $\begin{array}{r} 10 \\ -\ 3 \\ \hline \end{array}$	2. $\begin{array}{r} 7 \\ -\ 6 \\ \hline \end{array}$	3. $\begin{array}{r} 14 \\ -\ 7 \\ \hline \end{array}$	4. $\begin{array}{r} 8 \\ -\ 4 \\ \hline \end{array}$
5. $\begin{array}{r} 2 \\ -\ 0 \\ \hline \end{array}$	6. $\begin{array}{r} 17 \\ -\ 9 \\ \hline \end{array}$	7. $\begin{array}{r} 10 \\ -\ 5 \\ \hline \end{array}$	8. $\begin{array}{r} 5 \\ -\ 5 \\ \hline \end{array}$
9. $\begin{array}{r} 6 \\ -\ 1 \\ \hline \end{array}$	10. $\begin{array}{r} 11 \\ -\ 3 \\ \hline \end{array}$	11. $\begin{array}{r} 4 \\ -\ 2 \\ \hline \end{array}$	12. $\begin{array}{r} 12 \\ -\ 6 \\ \hline \end{array}$
13. $\begin{array}{r} 1 \\ -\ 0 \\ \hline \end{array}$	14. $\begin{array}{r} 3 \\ -\ 2 \\ \hline \end{array}$	15. $\begin{array}{r} 8 \\ -\ 4 \\ \hline \end{array}$	16. $\begin{array}{r} 15 \\ -\ 7 \\ \hline \end{array}$
17. $\begin{array}{r} 13 \\ -\ 5 \\ \hline \end{array}$	18. $\begin{array}{r} 8 \\ -\ 8 \\ \hline \end{array}$	19. $\begin{array}{r} 9 \\ -\ 3 \\ \hline \end{array}$	20. $\begin{array}{r} 10 \\ -\ 1 \\ \hline \end{array}$
21. $\begin{array}{r} 6 \\ -\ 2 \\ \hline \end{array}$	22. $\begin{array}{r} 3 \\ -\ 3 \\ \hline \end{array}$	23. $\begin{array}{r} 16 \\ -\ 9 \\ \hline \end{array}$	24. $\begin{array}{r} 11 \\ -\ 2 \\ \hline \end{array}$

Name ______________________________

My Review

Chapter 4
Subtraction Strategies to 20

Vocabulary Check

Vocab

Circle the correct answer.

1. count back

6 − 2 =

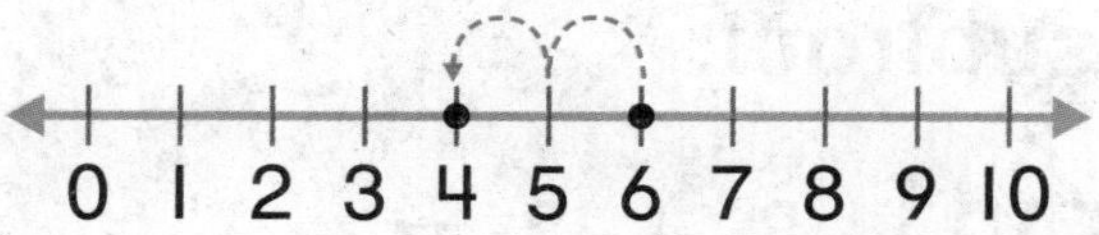

6 − 2 =

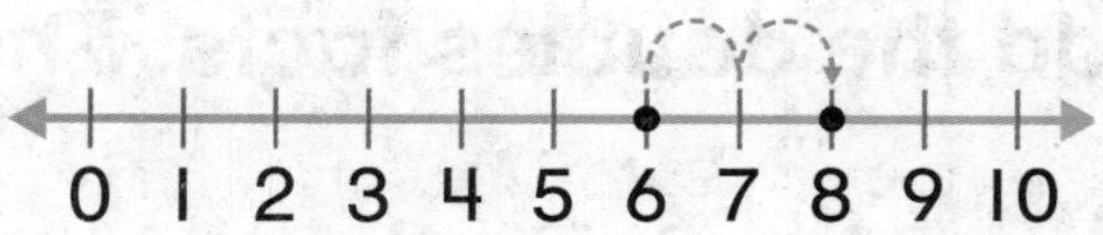

2. fact family

7 + 3 = 10
3 + 7 = 10
10 − 7 = 3
10 − 3 = 7

6 + 4 = 10
4 + 6 = 10
5 + 5 = 10
8 + 2 = 10

3. missing addend

8 + 3 = 11

8 + ☐ = 11

4. difference

↓
9 − 4 = 5

↓
6 − 1 = 5

Concept Check

Count back to subtract.

5. 7 − 3 = ____

6. 9 − 2 = ____

Use the number line to help you subtract.

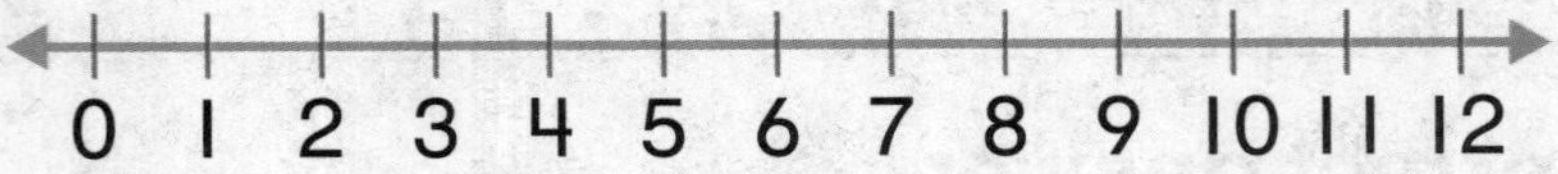

7. 11 − 2 = ____

8. 10 − 1 = ____

Add the doubles facts. Then subtract.

9. 6 + 6 = ____

12 − 6 = ____

10. 8 + 8 = ____

16 − 8 = ____

Add and subtract. Complete the fact family.

11.

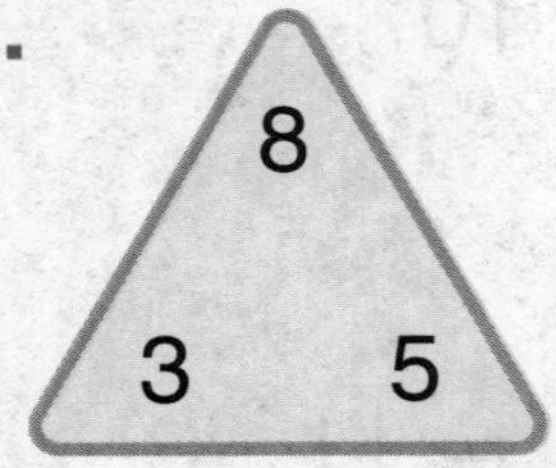

3 + 5 = ☐ 8 − 3 = ☐

5 + 3 = ☐ 8 − 5 = ☐

Find the missing addend.

12. 5 + ☐ = 12

12 − 5 = ☐

13. 9 + ☐ = 14

14 − 9 = ☐

Name ______________________

Problem Solving

14. Paige writes two related facts using these numbers. What related facts could she have written?

12, 8, 4

_____ + _____ = _____

_____ − _____ = _____

15. Jayson caught 15 fish. Ashlyn caught 7 fish. How many more fish did Jayson catch than Ashlyn?

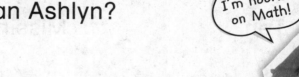

_____ − _____ = _____ fish

Test Practice

16. Cris has 14 jump ropes. What doubles fact shows the number of jump ropes Cris has?

7 + 8 = 14 jump ropes ○	7 + 7 = 14 jump ropes ○
14 − 5 = 9 jump ropes ○	6 + 6 = 12 jump ropes ○

Reflect

Chapter 4

Answering the Essential Question

Show the ways to subtract.

ESSENTIAL QUESTION

What strategies can I use to subtract?

Count Back

12, ____, ____, ____

12 − 3 = ____

Use Doubles

10 + 10 = ____

20 − 10 = ____

Make a Ten

17 − 8

☐ ☐

____ − ____ = ☐

____ − ____ = ☐

So, 17 − 8 = ____.

Missing Addend

5 + ☐ = 14

14 − 5 = ☐

Now I Know!

Chapter

5 Place Value

ESSENTIAL QUESTION
How can I use place value?

We're at the Toy Store!

Watch a video!

Name

Chapter 5 Project

Guess and Check

1. Trade bags with a different group.
2. Guess how many items are in the bag. Record that number on the chart.
3. Count to find how many items are in the bag. Record that number on the chart.
4. Put the items in groups of tens and ones. Then fill out the chart to show how many hundred(s), ten(s), and one(s) are in the total number of items from that bag.
5. Repeat the same process.

Guess	Number	Hundred(s)	Ten(s)	One(s)

Name

Am I Ready?

Circle to make groups of 10.

1.

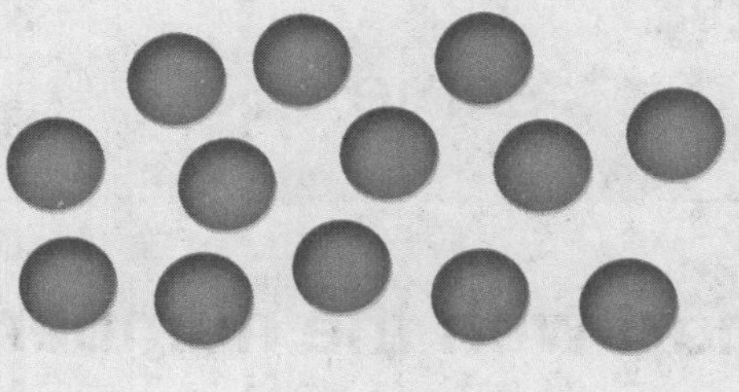

2.

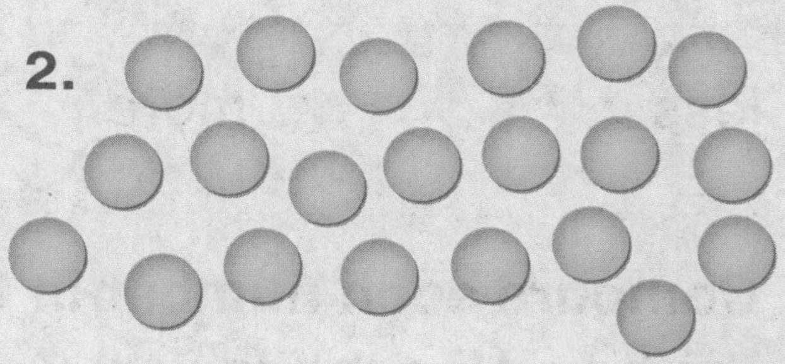

3. Write the missing numbers.

1	2	3		5	6	7	8	9	
11		13	14	15		17	18	19	20

4. Circle groups of 10. Count. Write the number.

_______ frogs

5. There are 4 boxes. Each box has 10 toys in it.
How many toys in all?

_______ toys

How Did I Do?

Shade the boxes to show the problems you answered correctly.

1	2	3	4	5

Name ....

My Math Words

Vocab

Review Vocabulary

less   more   same

Compare each number in the chart with the number in the box. Use the review words to complete the chart.

17

Name It!	Write It!	Draw It!
less	16	
same	17	
more	18	

My Vocabulary Cards

Vocab abc

Processes & Practices

Lesson 5-10

equal to (=)

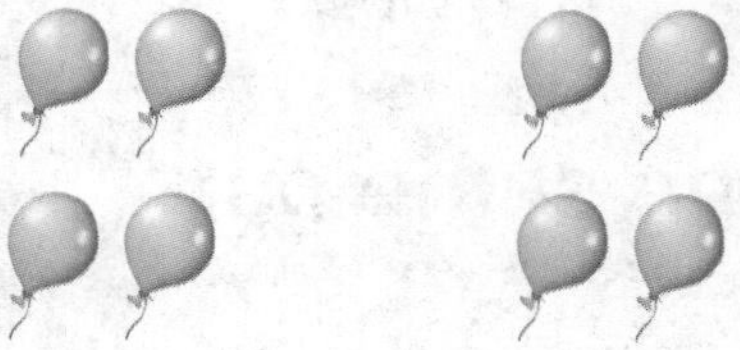

4 = 4

Lesson 5-10

greater than (>)

6 > 3

Lesson 5-12

hundred

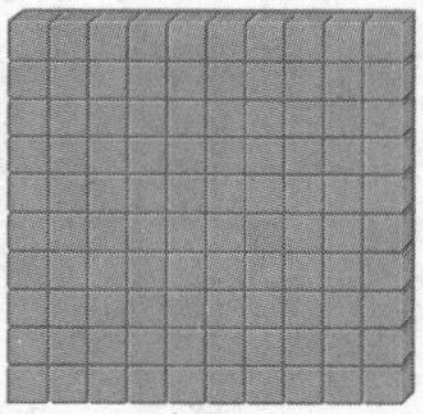

Lesson 5-10

less than (<)

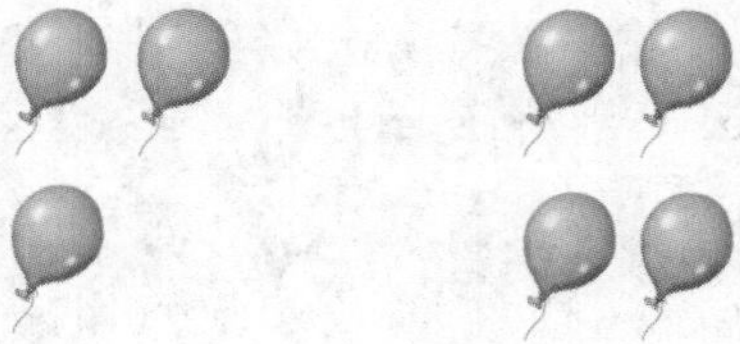

3 < 4

Lesson 5-5

ones

23

2 tens 3 ones

20 + 3

Lesson 5-5

regroup

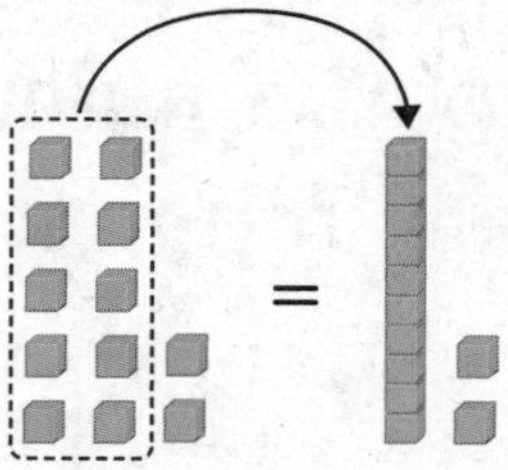

1 ten 2 ones

Teacher Directions:
Ideas for Use

- Tell students to create riddles for each word. Ask them to work with a friend to guess the word for each riddle.
- Have students draw examples for each card. Have them make drawings that are different from what is shown on each card.

The number or group with more.

Is the same as.

The number or group with fewer.

100 ones or 10 tens.

Take apart a number to write it in a new way.

The numbers in the range of 0 to 9.

My Vocabulary Cards

Vocab

Processes & Practices

Lesson 5-2

tens

23

2 tens 3 ones

20 + 3

Teacher Directions:
More Ideas for Use

- Ask students to use the blank cards to draw a picture that will help them with concepts such as groups of ten or more or ten more, ten less.
- Have students use the blank cards to write a word from a previous chapter that they would like to review.

The numbers in the range of 10 to 99.

My Foldable

FOLDABLES® Follow the steps on the back to make your Foldable.

0	1	2	3	4
5	6	7	8	9

tens	ones

FOLDABLES® Study Organizer

1
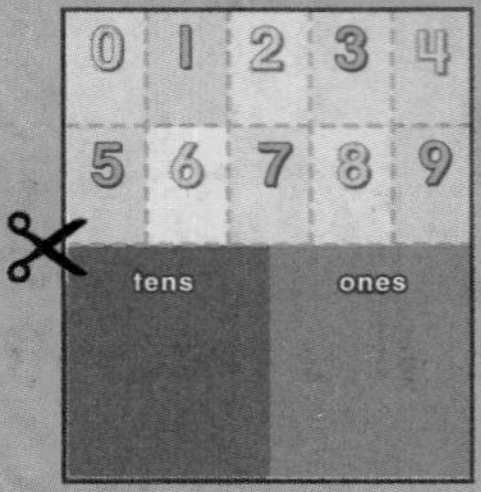

2
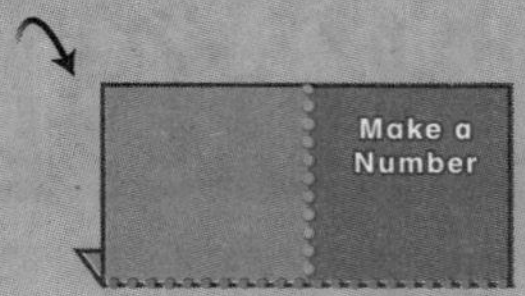

3

0	1	2	3	4
5	6	7	8	9

Make a Number

Name

Numbers 11 to 19

Lesson 1

ESSENTIAL QUESTION
How can I use place value?

Explore and Explain

Watch

10 and 4 more

Teacher Directions: Draw one toy in each box of the red toy bin. Color the toys red. Circle the group of 10 red toys. Draw 4 more toys inside the boxes of the yellow toy bin. Color these toys yellow. There are 10 and how many more toys? Trace your answer.

See and Show

Processes & Practices

Numbers from 11 to 19 can be made with one group of 10 and some more.

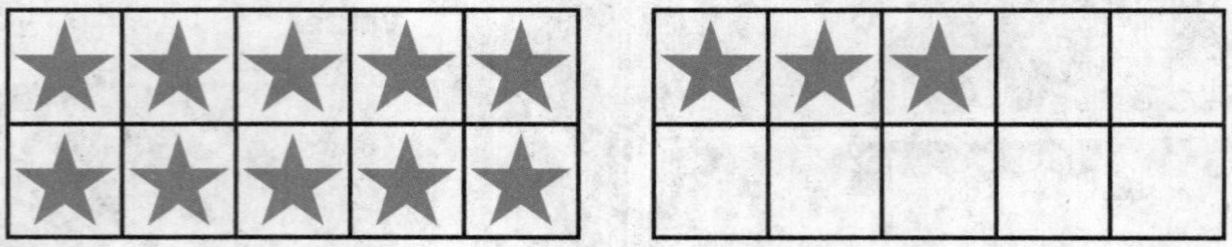

10 and 3 more

13 thirteen

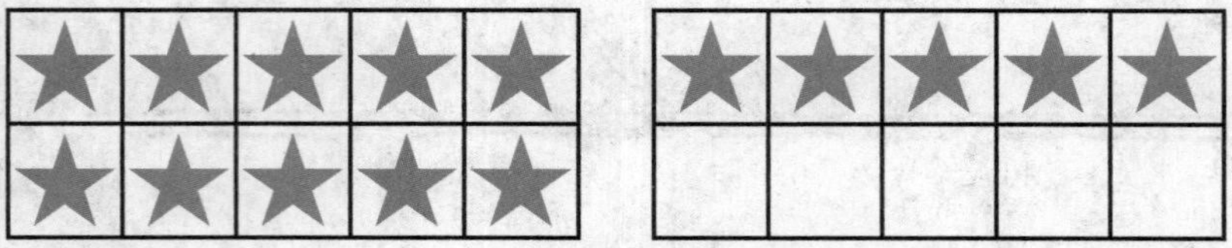

10 and 5 more

15 fifteen

Circle the group of ten. Write how many more. Then write the number.

11 eleven
12 twelve
13 thirteen
14 fourteen
15 fifteen

1. 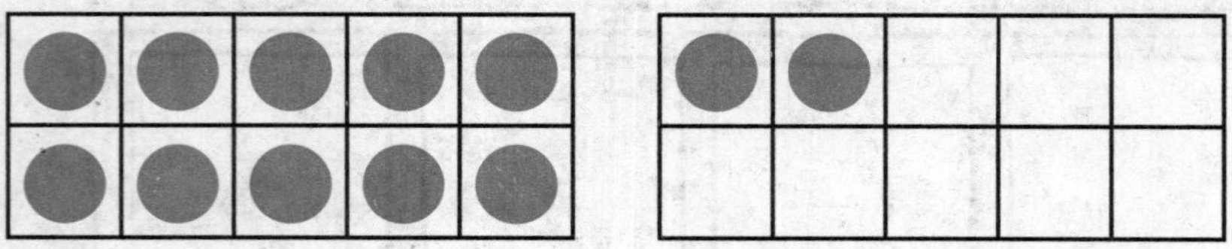

10 and _____ more

_____ twelve

2. 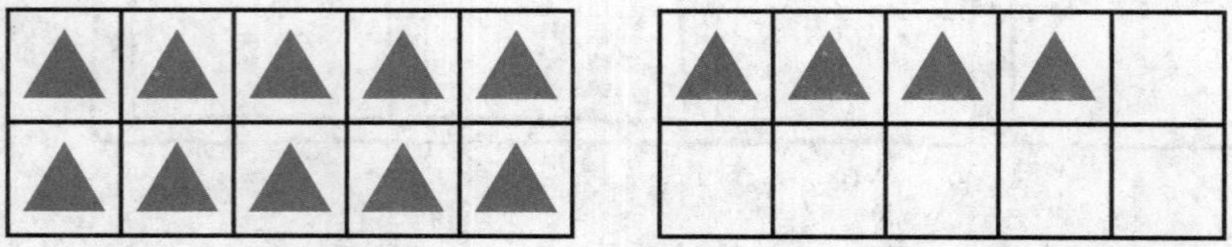

10 and _____ more

_____ fourteen

Talk Math How are these numbers alike: 11, 12, 13, 14, 15?

Name

On My Own

16 sixteen
17 seventeen
18 eighteen
19 nineteen

Circle the group of ten. Write how many more. Then write the number.

3. 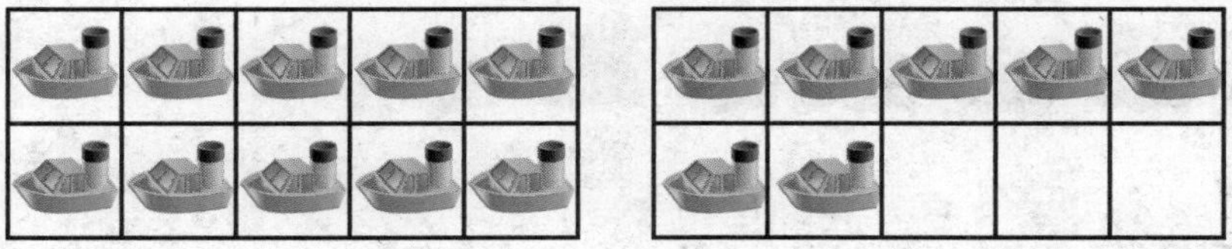

_____
seventeen

10 and _____ more

4. 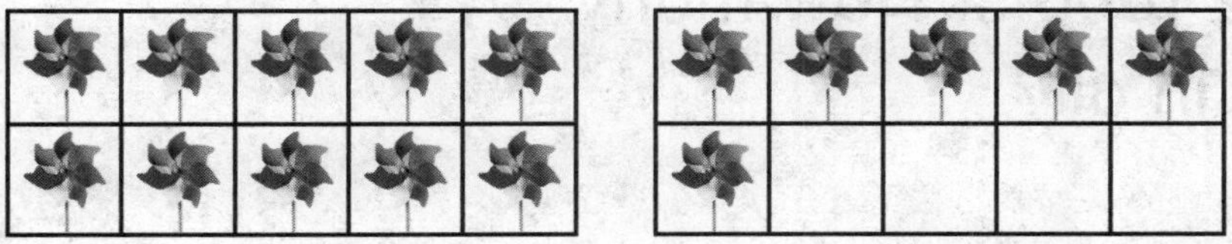

_____
sixteen

10 and _____ more

5. 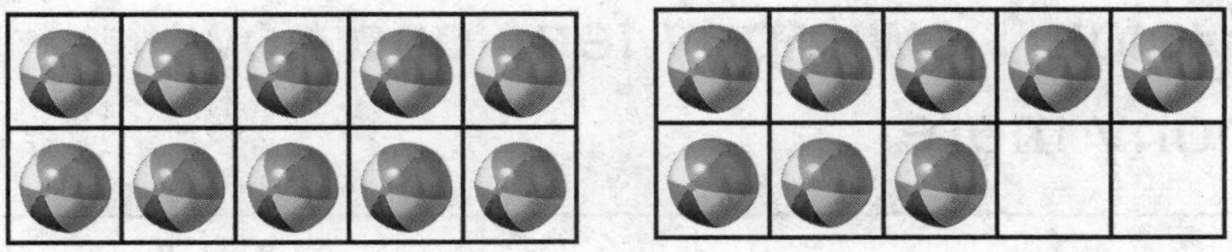

_____
eighteen

10 and _____ more

6. 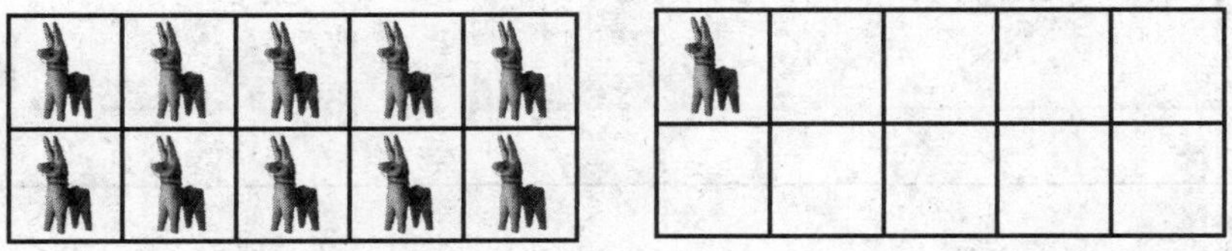

_____
eleven

10 and _____ more

Problem Solving

Processes & Practices

7. Ben has the cars shown. Circle a group of ten cars. Write how many more. How many cars does Ben have in all?

10 and _____ more _____ cars

8. Mara had 10 teddy bears in her room. She also had 2 robots. How many toys did Mara have in all?

_____ toys

Write Math Choose a number from 16 to 20. Explain how many tens and how many more.

Name ....

My Homework

Lesson 1

Numbers 11 to 19

Homework Helper

eHelp Need help? connectED.mcgraw-hill.com

The numbers from 11 to 19 can be made with one group of 10 and some more.

10 and 4 more

14
fourteen

Practice

Circle the group of ten. Write how many more. Then write the number.

1.

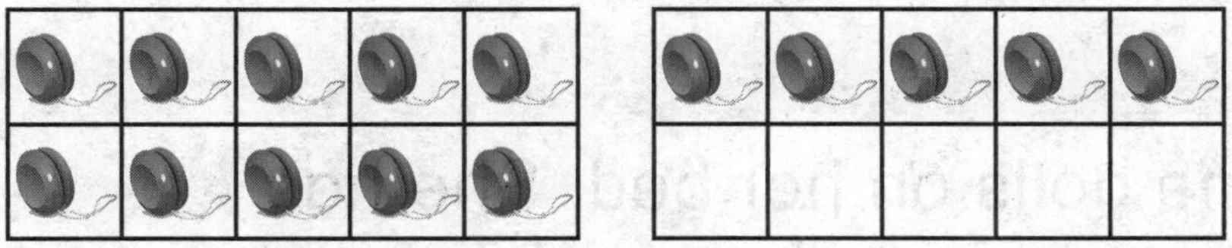

10 and ______ more

______
fifteen

2.

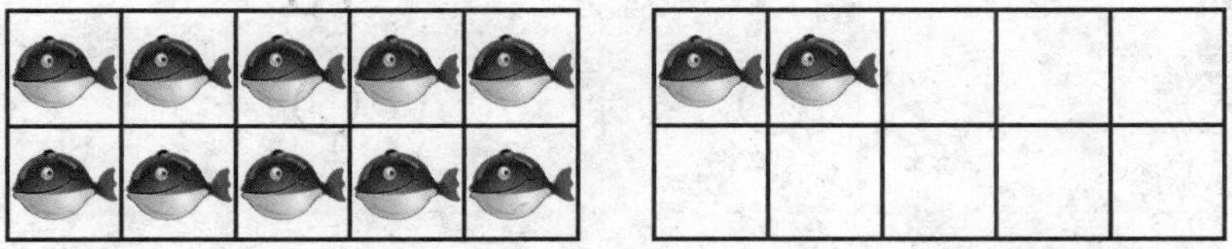

10 and ______ more

______
twelve

Circle the group of ten. Write how many more. Then write the number.

3. 

10 and ______ more

______ seventeen

4. 

10 and ______ more

______ eighteen

5. 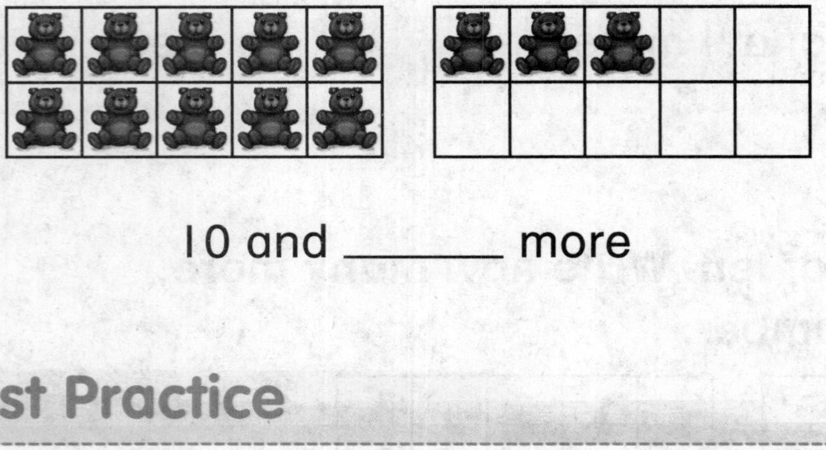

10 and ______ more

______ thirteen

Test Practice

6. Hanna has some dolls on her bed. She has 10 and 6 more. Which number shows how many dolls Hanna has on her bed?

4	6	10	16
○	○	○	○

Math at Home Say a number between 11 and 19. Have your child draw that many objects and write the number. Have him or her say how many groups of ten and how many more.

Name

Tens

Lesson 2

ESSENTIAL QUESTION
How can I use place value?

Explore and Explain

Watch Tools

Teacher Directions: Place one cube in each window of the train car. Join the cubes together to make a train. Move this train off the track. Repeat this two more times. Count by tens. Write the number.

See and Show

Processes & Practices

You can group 10 ones to make a **ten**.

 = 

3 tens = 30
thirty

Helpful Hint
These cubes are in groups of 10. Count by tens to find how many cubes there are in all.

Use . Make groups of ten. Count by tens. Write the numbers.

1. 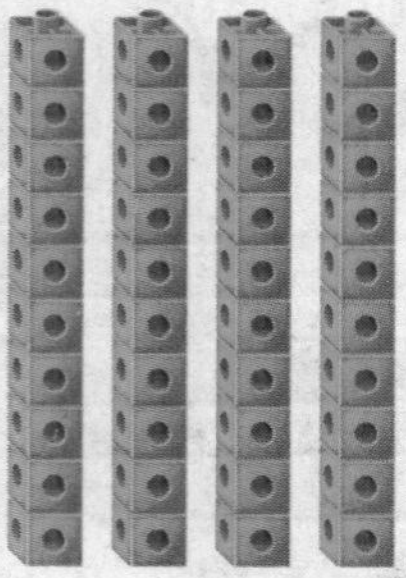

_____ tens = _____
forty

2. 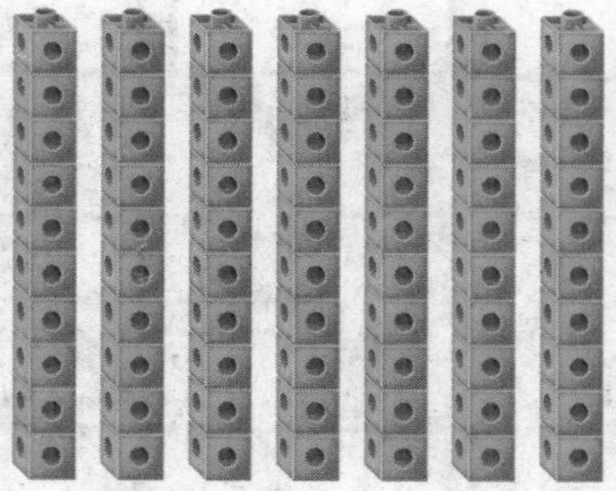

_____ tens = _____
seventy

Talk Math How would you use cubes to model the number 100?

Name ....

On My Own

Use [cube]. Make groups of ten. Write the numbers.

3. 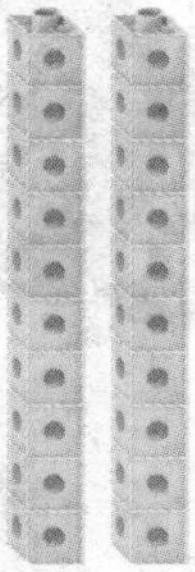

_______ tens = _______
twenty

4. 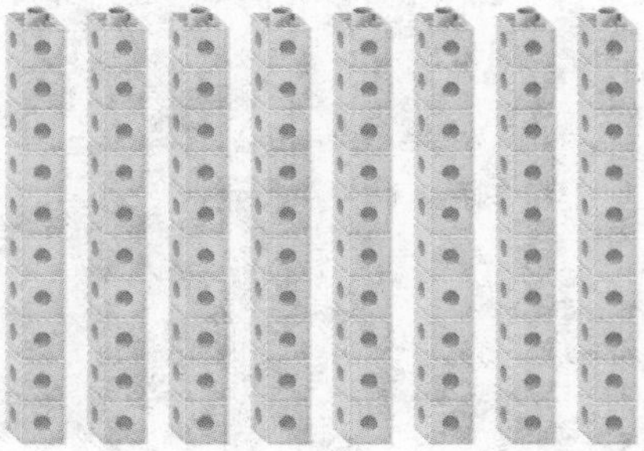

_______ tens = _______
eighty

5. 

_______ tens = _______
fifty

6. 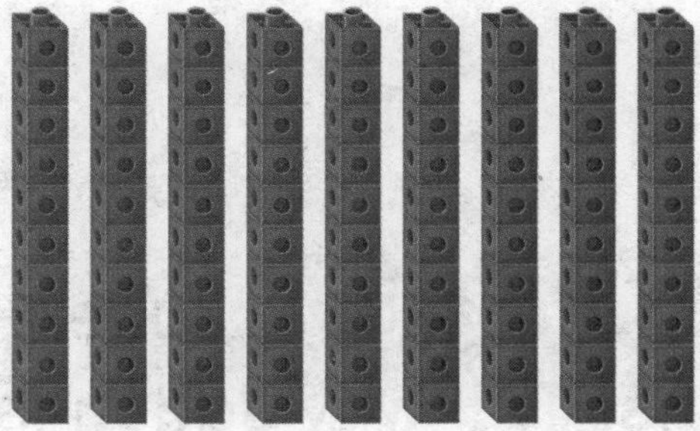

_______ tens = _______
ninety

7. 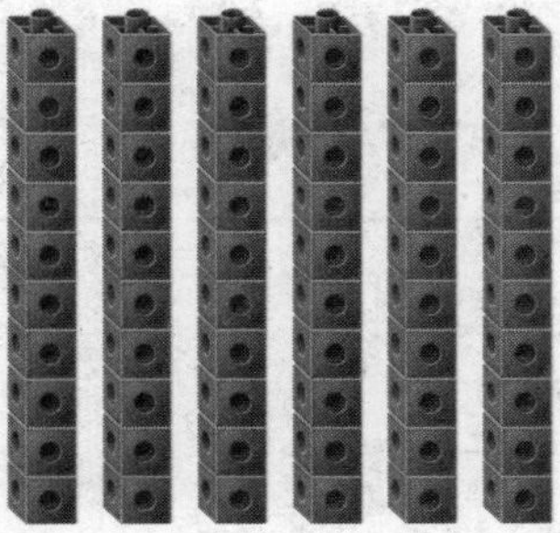

_______ tens = _______
sixty

8. 

_______ tens = _______
thirty

Problem Solving

Processes & Practices

9. Kelvin has 7 groups of cars.
There are 10 cars in each group.
How many cars does he have in all?

______ cars

10. Mrs. Brown is making cookies for the soccer players. There are 4 teams of 10.
How many cookies does she need?

______ cookies

HOT Problem Jaden says the pennies show 2 tens. Tell why Jaden is wrong. Make it right.

Name ____________________

My Homework

Lesson 2
Tens

Homework Helper

eHelp Need help? connectED.mcgraw-hill.com

Count by tens to find how many there are in all.

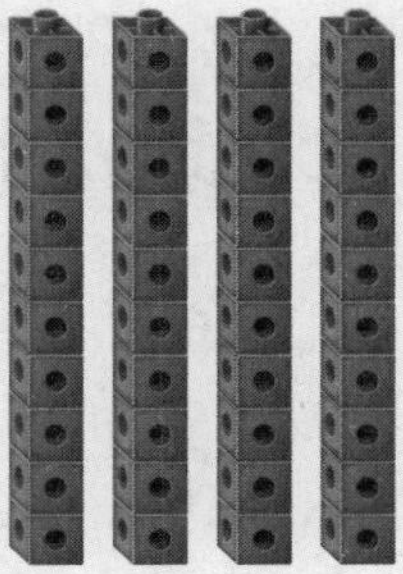

Helpful Hint
Count ten, twenty, thirty, forty.

4 tens = 40

Practice

Count groups of ten. Write the numbers.

1. 

_____ tens = _____
twenty

2. 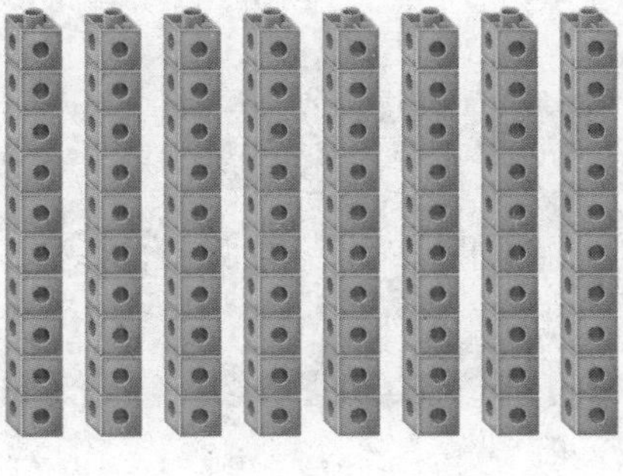

_____ tens = _____
eighty

Count groups of ten. Write the numbers.

3.

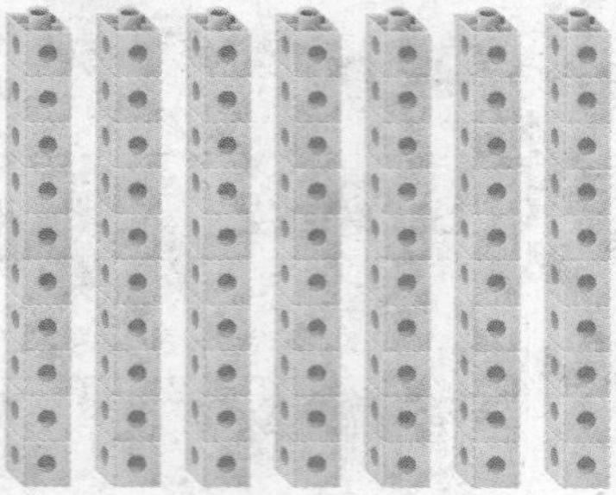

______ tens = ______
seventy

4.

______ tens = ______
fifty

5. There are 4 boxes. Each box has 10 buttons in it. How many buttons are there in all?

______ buttons

6. There are 3 vases. Each vase has 10 flowers in it. How many flowers are there in all?

______ flowers

Vocabulary Check

Circle the correct answer.

7. Which group shows counting by **tens**?

5, 10, 15, 20, 25   10, 20, 30, 40, 50

Math at Home Give your child several small items such as buttons or pennies to count. Help your child make groups of ten and then count the items by tens.

Name ....

Count by Tens Using Dimes

Lesson 3

ESSENTIAL QUESTION
How can I use place value?

Explore and Explain

Tools

_______ dimes   How much? _______¢

Teacher Directions: Circle all of the (dime) in the register. Write the number that shows the total number of dimes in the register. Count by tens to find how many cents (¢) are in the register. Write how much.

See and Show

Processes & Practices

You can count by tens to count dimes.

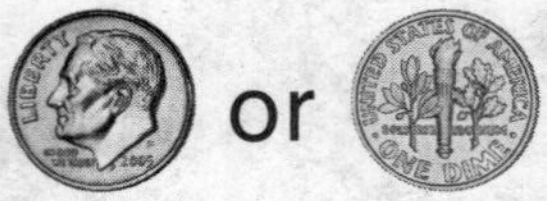

or

dime
10¢ = ten cents

Helpful Hint
10 pennies equal 1 dime.

10 ¢ 20 ¢ 30 ¢ 40 ¢ 50 ¢

Use (dime). Count by tens. Write the numbers. How much in all?

1.

____¢ ____¢ ____¢ in all

2.

____¢ ____¢ in all

3.

____¢ ____¢ ____¢ ____¢ ____¢ ____¢ ____¢ in all

Talk Math How many dimes are the same as 40 pennies?

Name ______________________

On My Own

Use (dime). Count by tens. Write the numbers. How much in all?

4.

_____¢ _____¢ _____¢ _____¢
in all

5.

_____¢ _____¢ _____¢ _____¢ _____¢ _____¢
in all

6.

_____¢ _____¢ _____¢ _____¢ _____¢
in all

7.

_____¢ _____¢ _____¢ _____¢ _____¢ _____¢ _____¢
in all

Problem Solving

Processes & Practices

8. You have 2 coins. They equal 20 cents. Draw the coins you have.

9. Kira is at the fair. She is playing a dime toss game. She throws 90¢ in dimes. Circle how many coins she tosses.

HOT Problem Kieran has 2 dimes, Austin has 4 dimes, and Leroy has 1 dime. How much money do they have in all? Explain.

Name ____________________

My Homework

Lesson 3

Count by Tens Using Dimes

Homework Helper

eHelp

Need help? connectED.mcgraw-hill.com

You can count by tens to count dimes.

 or  = 10¢

10¢ 20¢ 30¢ 40¢ 50¢ 60¢ in all

Practice

Count by tens. Write the numbers. How much in all?

1.

_____¢ _____¢ _____¢
in all

2.

_____¢ _____¢ _____¢ _____¢ _____¢ _____¢ _____¢
in all

Count by tens. Write the numbers. How much in all?

3.

_____¢ _____¢ _____¢ _____¢ _____¢

in all

4.

_____¢ _____¢ _____¢ _____¢ _____¢ _____¢

in all

5. Addison has 8 dimes in her purse.
How much is there in all?

_____ ¢

Test Practice

6. Myron buys a small basketball for 90¢.
How many dimes show 90¢?

10 dimes ○ 9 dimes ○ 8 dimes ○ 7 dimes ○

Math at Home Give your child different amounts of dimes from 1 to 9. Have them tell you how many dimes they have and how much money in all.

Name ______________________________

Ten and Some More

Lesson 4

ESSENTIAL QUESTION
How can I use place value?

Explore and Explain

Watch Tools

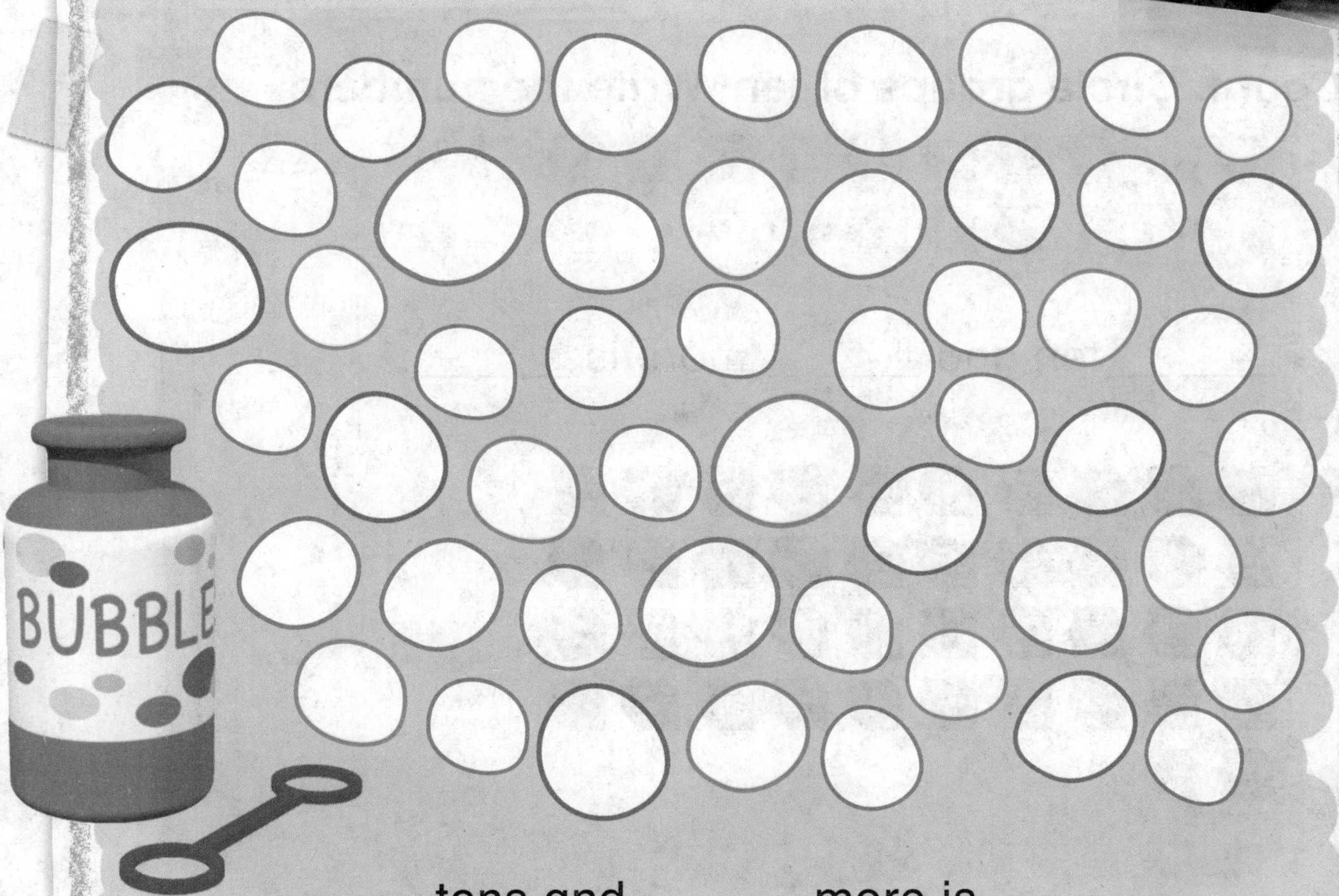

______ tens and ______ more is ______.

Teacher Directions: Use crayons to color groups of ten. Make each group a different color. Write the numbers.

See and Show

You can count groups of ten and some more.

Helpful Hint
There are two groups of ten. There are 9 more cubes.

__2__ tens and __9__ more is __29__.

Count. Circle groups of ten. Write the numbers.

1. 

_____ ten and _____ more is _____.

2. 

_____ tens and _____ more is _____.

Talk Math What number is 7 tens and 0 more? How do you know?

Name ....

On My Own

Count. Circle groups of ten. Write the numbers.

3. 

_____ ten and _____ more is _____.

4. 

_____ tens and _____ more is _____.

5. 

_____ tens and _____ more is _____.

6. 

_____ tens and _____ more is _____.

Problem Solving

Processes & Practices

7. Julia has 4 groups of 10 bracelets. She also has 3 more bracelets. How many bracelets does she have in all?

_____ bracelets

8. Landon has 5 groups of 10 baseball cards. He also has 7 more baseball cards. How many baseball cards does he have in all?

_____ baseball cards

HOT Problem Morgan wrote this sentence. Tell why Morgan is wrong. Make it right.

3 tens and 6 more is 16.

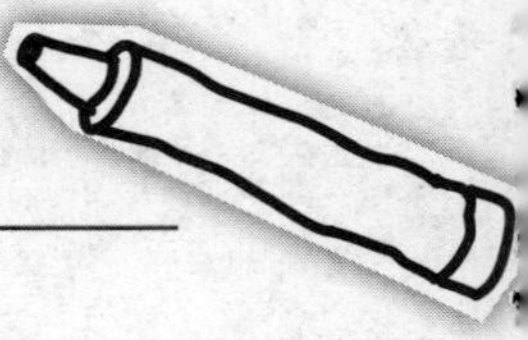

____

____

____

Name ______________________________

My Homework

Lesson 4
Ten and Some More

Homework Helper

eHelp **Need help?** connectED.mcgraw-hill.com

You can count groups of ten and some more.

2 tens and 3 more is 23.

Helpful Hint
Count tens and then count ones.

Practice

Count. Circle groups of ten. Write the numbers.

1.

_____ tens and _____ more is _____.

2.

_____ ten and _____ more is _____.

Count. Circle groups of ten. Write the numbers.

3. 

_____ tens and _____ more is _____.

4. 

_____ tens and _____ more is _____.

5. Sarai has 30 rings. She gets 2 more rings. How many rings does Sarai have?

_____ rings

Test Practice

6. Paul is thinking of a number. It has 6 groups of ten and 3 more. What is the number?

63	36	9	3
○	○	○	○

Math at Home Have your child count some cereal by ones. Then make groups of ten and some more. Ask your child how many pieces of cereal in all.

Name

Tens and Ones

Lesson 5

ESSENTIAL QUESTION
How can I use place value?

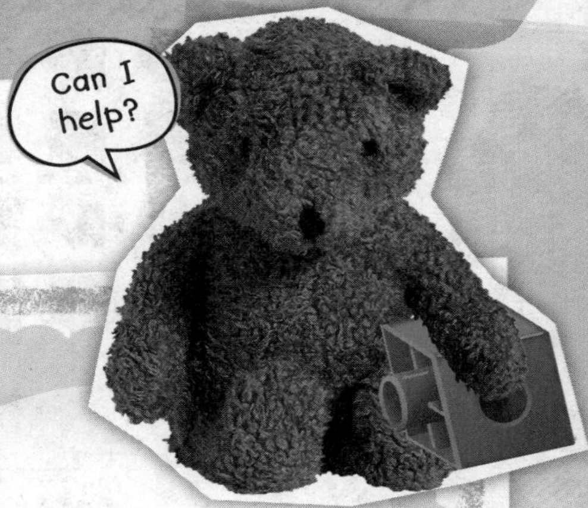

Explore and Explain

Watch Tools

23 ones = ______ tens and ______ ones = ______

Teacher Directions: Use cubes to show 23 ones. Connect the cubes to make groups of 10 and some more. Trace the cubes. Write how many groups of tens and ones. Then write the number.

See and Show

Processes & Practices

You can show a number as tens and ones.
You can **regroup** 10 **ones** as 1 ten.

Helpful Hint
Put together 10 ones to make 1 ten.

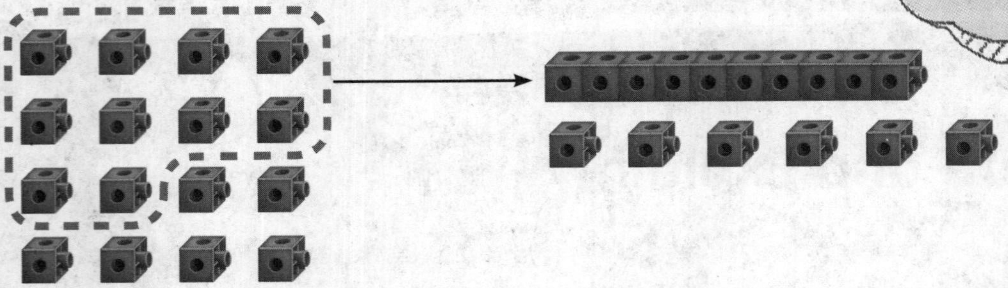

16 ones = __1__ ten and __6__ ones = __16__

Use . Circle groups of ten. Write how many tens and ones. Write how many there are in all.

1.

14 ones = ______ ten and ______ ones = ______

2.

21 ones = ______ tens and ______ one = ______

Talk Math How would you regroup 30 ones?

Name ______________________________

On My Own

Use [cube]. Circle groups of ten. Write how many tens and ones. Write how many there are in all.

3.

13 ones = ______ ten and ______ ones = ______

4.

26 ones = ______ tens and ______ ones = ______

Make groups of tens and ones. Write how many.

5. 35 ones = ______ tens and ______ ones = ______

6. 47 ones = ______ tens and ______ ones = ______

7. 67 ones = ______ tens and ______ ones = ______

8. 29 ones = ______ tens and ______ ones = ______

Problem Solving

Processes & Practices

9. There are 24 ones. Write how many tens and ones. Write how many there are in all.

24 ones = ______ tens and ______ ones = ______

10. Abby has 45 stickers. She has 4 sets of 10 stickers that are flowers. The rest are butterflies. How many of the stickers are butterflies?

______ stickers

Write Math Explain how to regroup 51 ones as tens and ones.

Name ______________________________

My Homework

Lesson 5
Tens and Ones

Homework Helper

eHelp Need help? connectED.mcgraw-hill.com

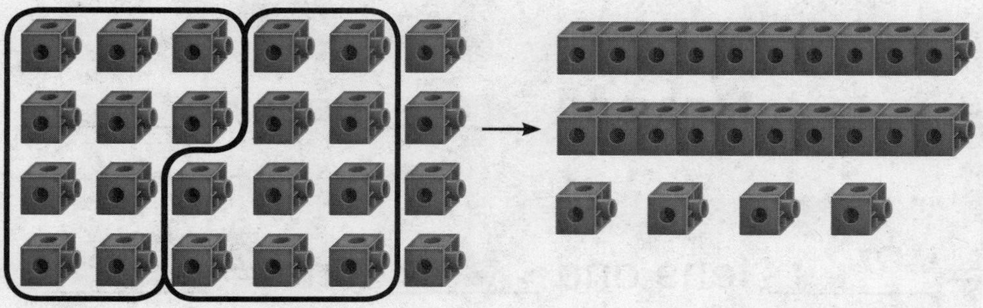

Helpful Hint
Put together 10 ones to make 1 ten.

24 ones = 2 tens and 4 ones = 24

Practice

Circle groups of ten. Write how many tens and ones. Write how many there are in all.

1.

15 ones = _____ ten and _____ ones = _____

2.

32 ones = _____ tens and _____ ones = _____

Make groups of tens and ones. Write how many.

3. 52 ones = ______ tens and ______ ones = ______

4. 77 ones = ______ tens and ______ ones = ______

5. 55 ones = ______ tens and ______ ones = ______

6. 80 ones = ______ tens and ______ ones = ______

7. 91 ones = ______ tens and ______ one = ______

8. Maria has 20 pencils in one box. She has 3 pencils in another box. How many pencils does Maria have in all?

______ pencils

Vocabulary Check

Vocab

Write the missing word.

ones **regroup**

9. You can ____________ 10 ones as 1 ten.

Math at Home Ask your child to regroup 83 as tens and ones.

Name ....

Check My Progress

Vocabulary Check

Complete the sentences.

regroup **tens**

1. The 2 in the number 25 shows the ______________.

2. You can ______________ by taking apart a number and writing it in a new way.

Concept Check

Circle the group of ten. Write how many more. Then write the number.

3. 

10 and _____ more

_______ eighteen

4. 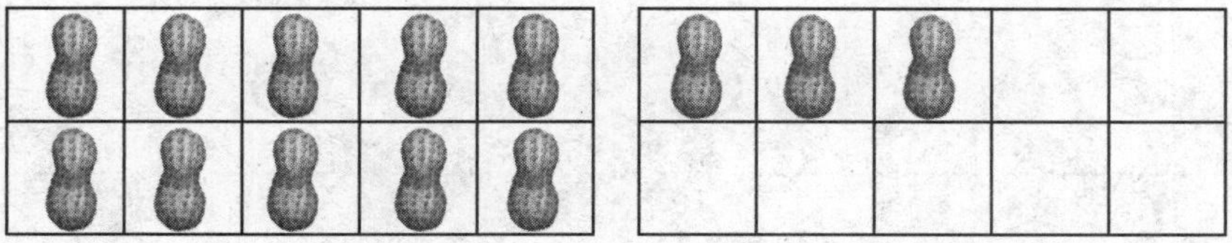

10 and _____ more

_______ thirteen

Count groups of ten. Write the numbers.

5.

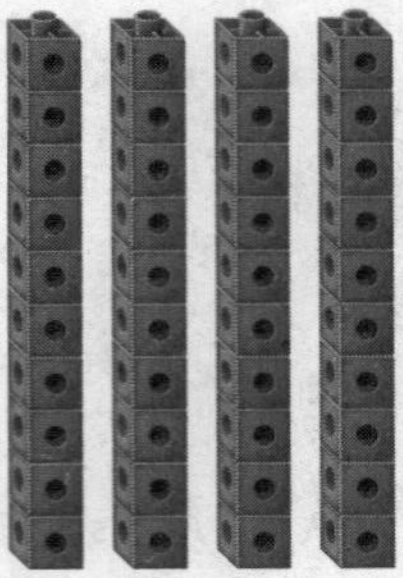

_____ tens = _____
forty

6.

_____ tens = _____
thirty

Count. Circle groups of ten. Write the numbers.

7.

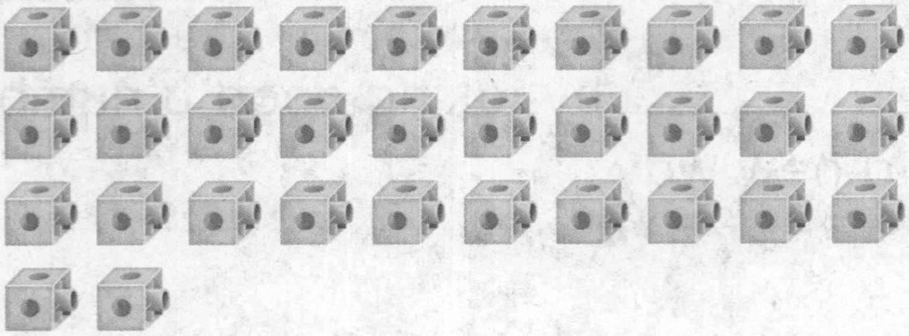

_____ tens and _____ more is _____.

Test Practice

8. Suzi is thinking of a number. It has 9 tens and 7 ones. What is the number?

70	79	90	97
○	○	○	○

Name ....

Problem Solving

STRATEGY: Make a Table

Lesson 6

ESSENTIAL QUESTION
How can I use place value?

Watch

Joel drinks 10 glasses of milk each week. How many glasses of milk does he drink in 4 weeks?

1 Understand Underline what you know.
Circle what you need to find.

2 Plan How will I solve the problem?

3 Solve I will make a table.

4 tens = 40
forty

40 glasses

Week	Glasses
1	10
2	20
3	30
4	40

4 Check Is my answer reasonable? Explain.

Practice the Strategy

Lea played with 10 different toys each day. How many toys would she have played with in 5 days?

1 Understand Underline what you know.
Circle what you need to find.

2 Plan How will I solve the problem?

3 Solve I will...

Day	Toys

_______ tens = _______

_______ toys

4 Check Is my answer reasonable? Explain.

Name ....

Processes & Practices

Apply the Strategy

1. Bobby found shells at the beach. He found 10 shells each day for 6 days. How many shells did he find in 6 days?

Day	Shells

_____ shells

2. A doctor sees 10 children a day. How many children does he see in 4 days?

Day	Children

_____ children

3. 10 birds fly south every day. How many birds will fly south after 7 days?

Day	Birds

_____ birds

Review the Strategies

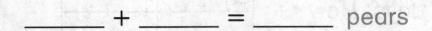

Choose a strategy

- Make a table.
- Act it out.
- Write a number sentence.

4. Brianna puts 10 pears into each of 2 bowls. How many pears are there in all?

_____ + _____ = _____ pears

5. Ms. Kim has 18 balloons. She gives Rae and Doug the same number of balloons. How many balloons does she give each person?

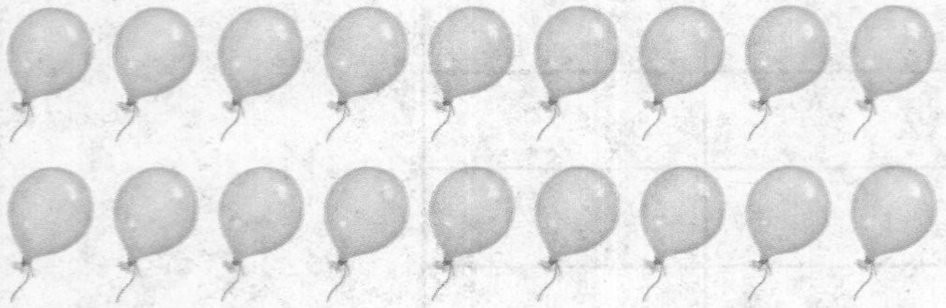

_____ balloons

6. Hill Elementary has 4 spelling bee teams. Each team has 10 students. How many students are there in all?

Team	Students
1	
2	
3	
4	

_____ students

Name ....

My Homework

Lesson 6

Problem Solving: Make a Table

Homework Helper

eHelp Need help? connectED.mcgraw-hill.com

Dave puts 10 stuffed toys in each box. There are 6 boxes. How many stuffed toys does Dave have?

1 Understand Underline what you know. Circle what you need to find.

2 Plan How will I solve the problem?

3 Solve I will make a table.

6 tens = 60

60 toys

Box	Toys
1	10
2	20
3	30
4	40
5	50
6	60

4 Check Is my answer reasonable?

Problem Solving

1. Mara has 3 groups of 10 balls. How many balls does Mara have in all?

Group	Balls

_____ balls

2. There are 10 ducks in each pond. How many ducks are there in 4 ponds?

Pond	Ducks

_____ ducks

3. Toy whistles are sold in bags of 10. Alexis needs 50 whistles. How many bags should she buy?

Bags	Whistles

_____ bags

Math at Home Take advantage of problem-solving opportunities during daily routines such as riding in the car, bedtime, doing laundry, putting away groceries, planning schedules and so on.

Name ____________________

Numbers to 100

Lesson 7

ESSENTIAL QUESTION
How can I use place value?

What's your favorite game?

Explore and Explain

Watch
Tools

tens	ones

____
thirty-eight

Teacher Directions: Use [ten-rod] and [unit cube]. Show 38 in groups of tens and ones. Write the number.

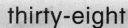

See and Show

Processes & Practices

You can write numbers in different ways.

tens	ones

tens	ones
4	2

4 tens 2 ones

42 forty-two

Use Work Mat 7 and  and . Show groups of tens and ones. Write the tens and ones. Then write the number.

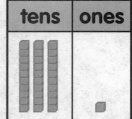

1.

tens	ones
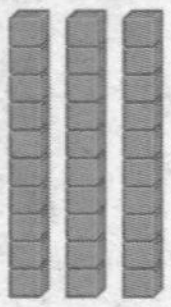	

tens	ones

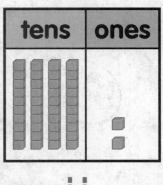

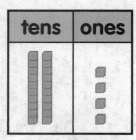

_____ thirty-one

2.

tens	ones
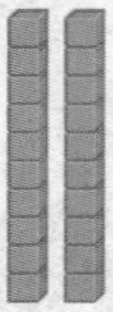	

tens	ones

_____ twenty-four

Talk Math How can you write 72 in more than one way?

Name ....

On My Own

Use Work Mat 7 and [ten rod] and [unit cube]. Show groups of tens and ones. Write the tens and ones. Then write the number.

3.

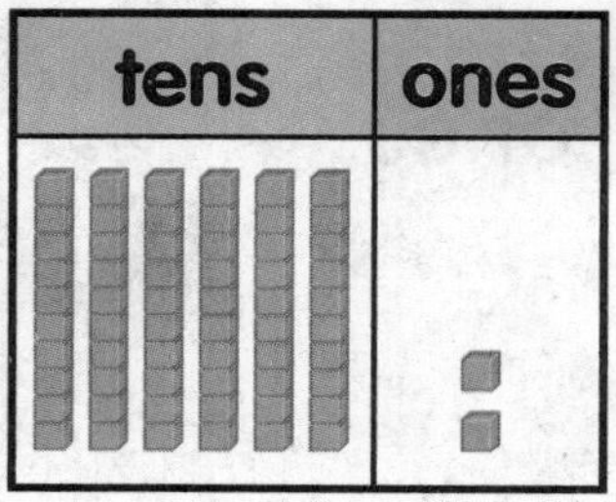

tens	ones

______ sixty-two

4.

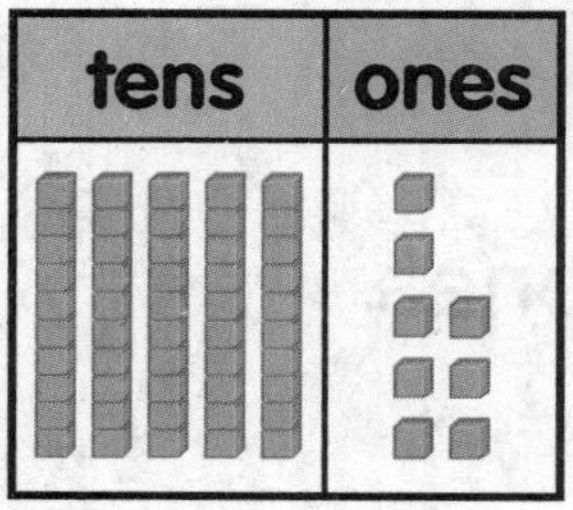

tens	ones

______ fifty-eight

5.

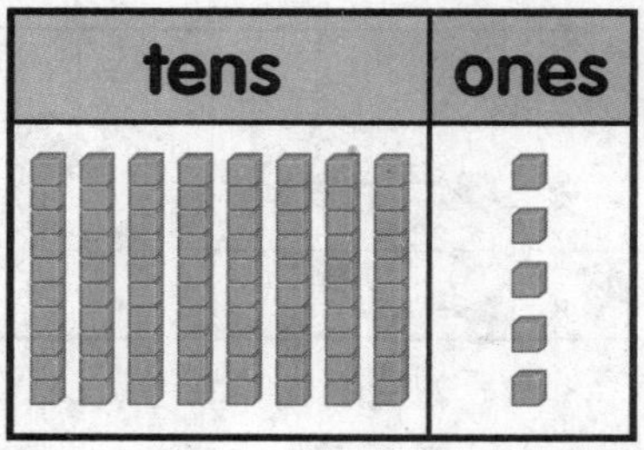

tens	ones

______ eighty-five

6.

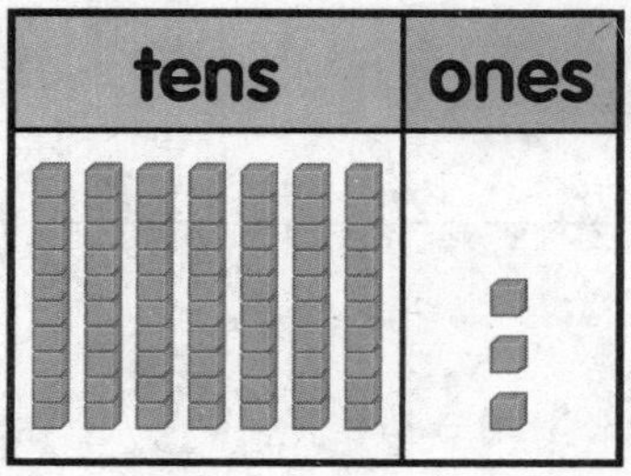

tens	ones

______ seventy-three

Problem Solving

Processes & Practices

7. Edgar has 47 blocks. How many groups of ten does he have? How many ones?

______ tens and ______ ones

8. Evan found 34 leaves in the park. How many groups of 10 does he have? How many ones?

______ tens and ______ ones

Write Math Explain how to show 84 using tens and some ones.

Name ________________________

My Homework

Lesson 7
Numbers to 100

Homework Helper

Need help? connectED.mcgraw-hill.com

You can write numbers in different ways.

tens	ones

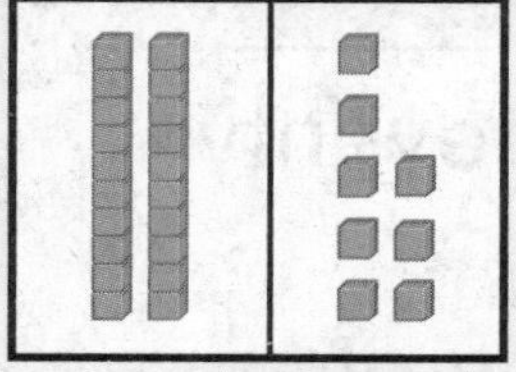

tens	ones
2	8

28
twenty-eight

2 tens 8 ones

Practice

Count groups of tens and ones. Write the tens and ones. Then write the number.

1.

tens	ones

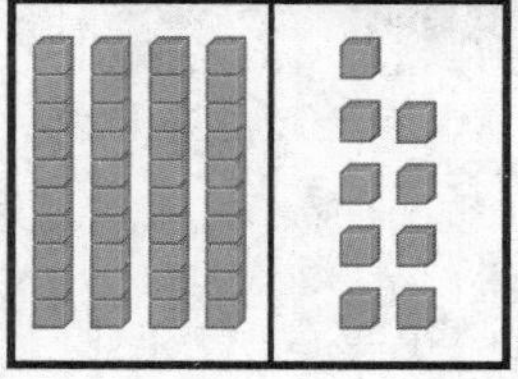

tens	ones

______
forty-nine

2.

tens	ones

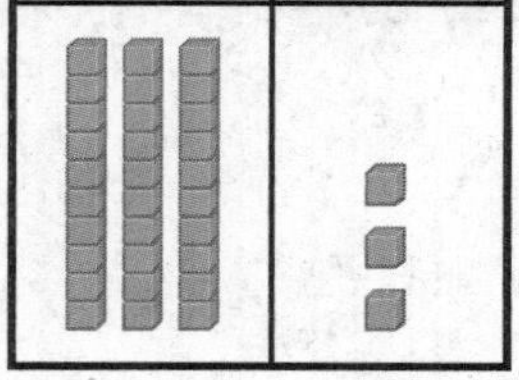

tens	ones

______
thirty-three

Count groups of tens and ones. Write the tens and ones. Then write the number.

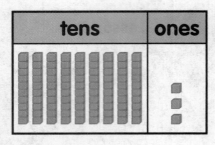

3.

tens	ones
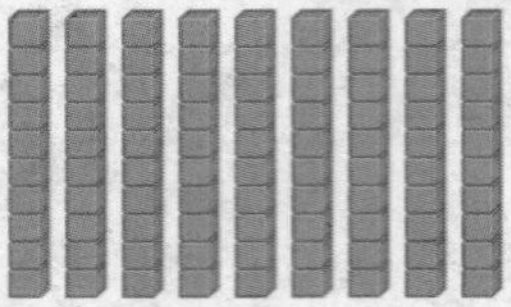	

tens	ones

______
ninety-three

4.

tens	ones
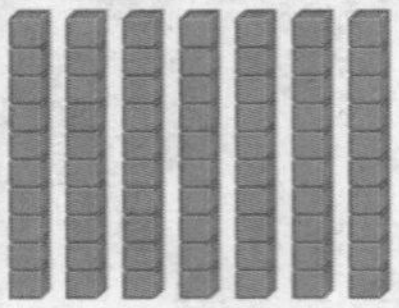	

tens	ones

______
seventy

5. Scarlet has 23 dolls. How many groups of ten does she have? How many ones?

______ tens and ______ ones

Test Practice

6. How many tens and ones are in 58?

5 tens 9 ones ○	4 tens 8 ones ○
4 tens 9 ones ○	5 tens 8 ones ○

Math at Home Ask your child to show 64 in two different ways.

Name

Ten More, Ten Less

Lesson 8

ESSENTIAL QUESTION
How can I use place value?

Let's peel out of here!

Explore and Explain

33
43
53

53 is ten more than 43.

33 is ten less than 43.

Teacher Directions: The monkey is pointing to the number 43. Place a counter on the number. Circle the number that is 10 more in blue. Trace it on the line below. Circle the number that is 10 less in red. Trace it on the line below.

See and Show

Processes &Practices

You can use mental math to find ten more or ten less than a number.

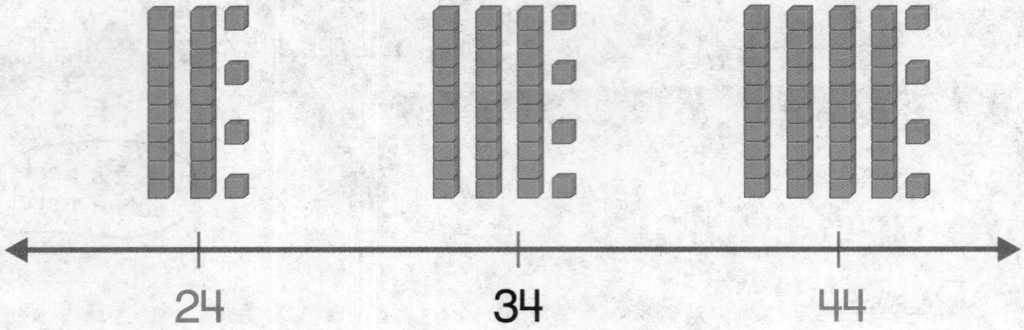

44 is ten more than 34.

24 is ten less than 34.

Write the missing number.

1.

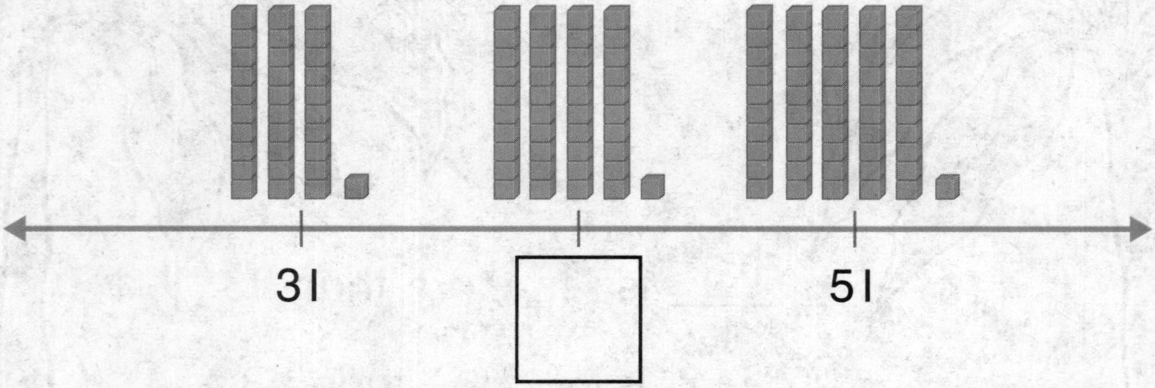

Talk Math Tell how to find what number is ten more than 62.

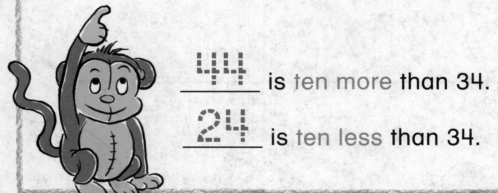

Name

On My Own

Write the missing number.

2.

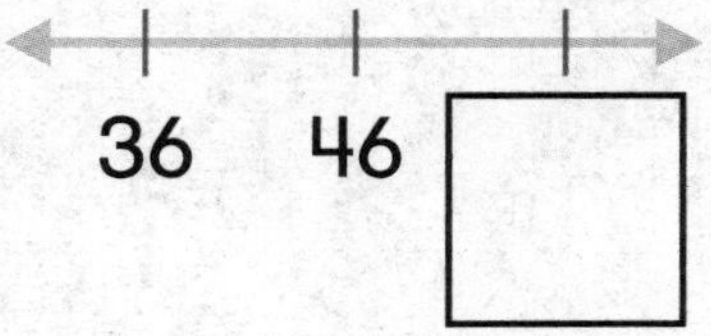

3.

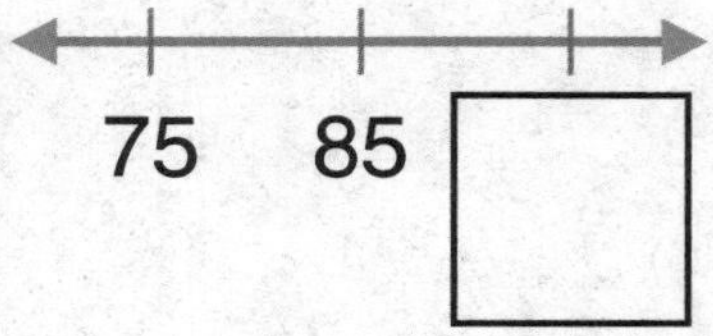

4.

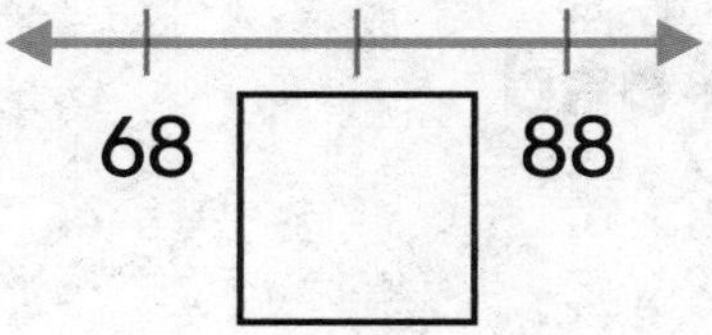

5.

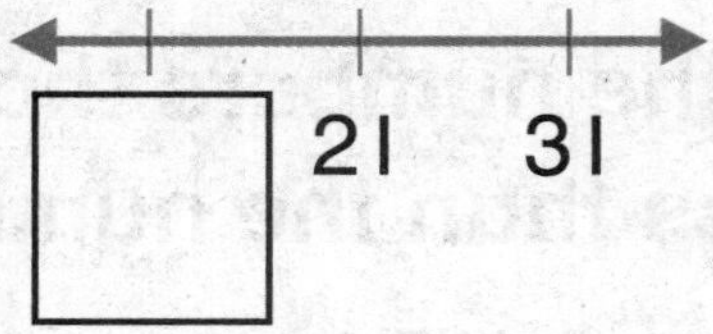

Write the number that is ten more.

6. 0, ______ **7.** 89, ______ **8.** 14, ______

9. 72, ______ **10.** 28, ______ **11.** 55, ______

Write the number that is ten less.

12. ______, 100 **13.** ______, 73 **14.** ______, 91

15. ______, 47 **16.** ______, 36 **17.** ______, 68

18. What number is ten more than 34?

______

Problem Solving

Processes & Practices

19. Jack had 29 toy cars. He was given 10 more toy cars at his party. How many toy cars does he have now?

_____ toy cars

Write the numbers that are ten more and ten less than the number.

20. _____, 31, _____

21. _____, 79, _____

22. _____, 24, _____

23. _____, 88, _____

HOT Problem Hailey had 50 pennies. Logan had ten less pennies. The answer is 40 pennies. What is the question?

Name ....

My Homework

Lesson 8
Ten More, Ten Less

Homework Helper

eHelp **Need help?** connectED.mcgraw-hill.com

You can use mental math to find ten more or ten less than a number.

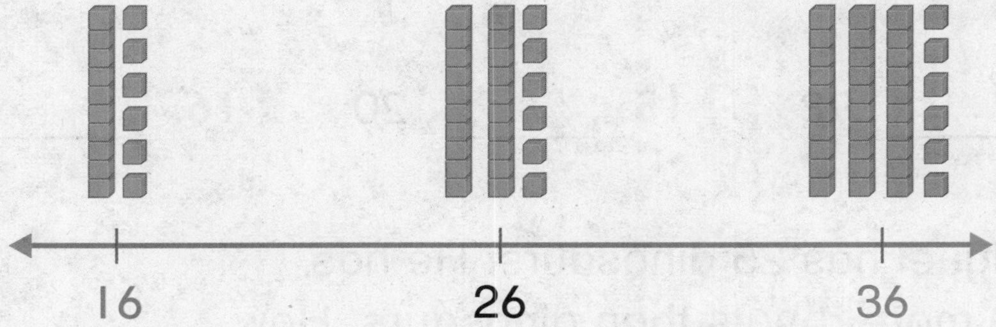

36 is ten more than 26.
16 is ten less than 26.

Helpful Hint
Count by 10s.
16, 26, 36

Write the missing number.

1. ☐ 15 25

2. ☐ 66 76

3. 73 ☐ 93

4. 43 53 ☐

Write the number that is ten more.

5. 12, ______
6. 20, ______
7. 77, ______
8. 9, ______
9. 26, ______
10. 55, ______

Write the number that is ten less.

11. ______, 57
12. ______, 31
13. ______, 82
14. ______, 98
15. ______, 20
16. ______, 71

17. Miguel has 25 dinosaurs. He has 10 more robots than dinosaurs. How many robots does Miguel have?

______ robots

Test Practice

18. Rhonda has 24 books. She gets 10 more books. How many books does Rhonda have in all?

70	55	34	18
○	○	○	○

Math at Home Write a multiple of ten such as 10, 20, 30, 40, 50, etc. up to 80 on a piece of paper. Ask your child to write the number that is ten more or ten less than that number.

Name ______________________

Count by Fives Using Nickels

Lesson 9

ESSENTIAL QUESTION
How can I use place value?

Explore and Explain

Tools

or nickel = 5¢

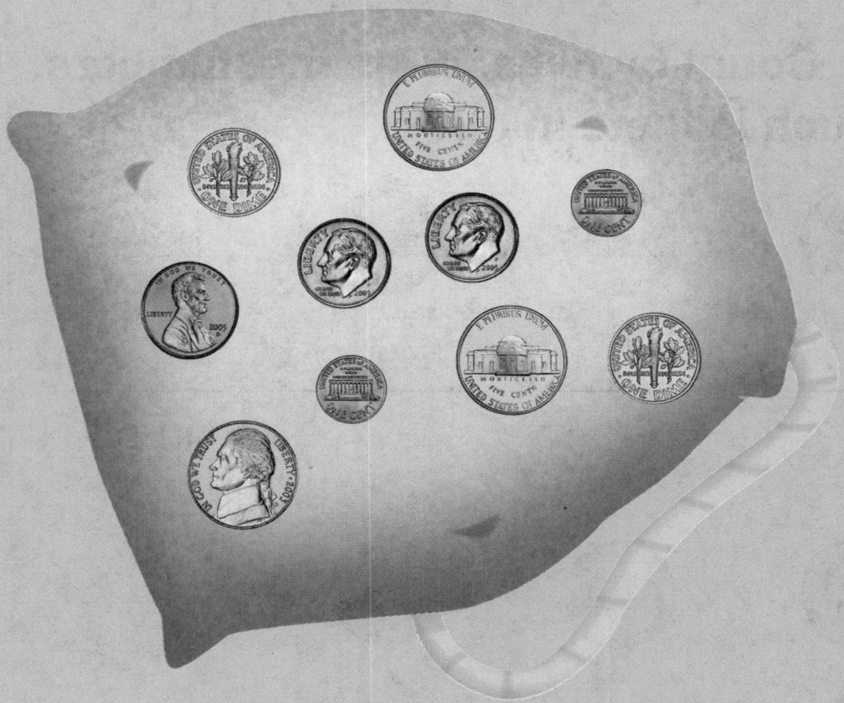

______ nickels How much? ______¢

Teacher Directions: Circle all of the nickels in the bag. Write the number that shows the total number of nickels in the bag. Count by fives to find how many cents (¢) are in the bag. Write how much there is in the bag.

See and Show

Processes & Practices

You can count by fives to count nickels.

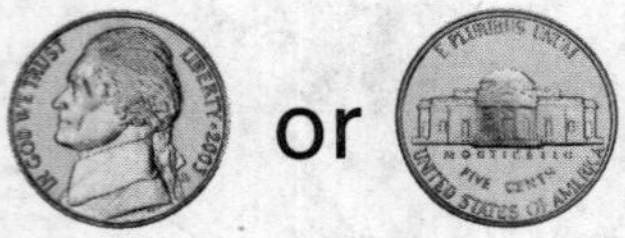

or

nickel
5¢ = five cents

Helpful Hint
5 pennies equal 1 nickel, so you could trade 5 pennies for 1 nickel.

5¢ 10¢ 15¢ 20¢ 25¢

Use (nickel). Count by fives. Write the numbers. How much is there in all?

1.

____¢ ____¢ ____¢ ____¢ in all

2.

____¢ ____¢ ____¢ in all

Talk Math My friend wants to give me 1 nickel for 10 pennies. Is that a fair trade? Explain.

Name ______________________________

On My Own

Use . Count by fives. Write the numbers.
How much is there in all?

3.

______ ¢ ______ ¢
in all

4.

______ ¢ ______ ¢ ______ ¢ ______ ¢ ______ ¢

______ ¢ ______ ¢
in all

5.

______ ¢ ______ ¢ ______ ¢ ______ ¢ ______ ¢

______ ¢ ______ ¢ ______ ¢
in all

Problem Solving

Processes & Practices

Draw a picture to solve.

6. Mindy has 11 nickels. How much money does she have?

_____ ¢

7. Ben buys a toy that costs 30¢. The only coins he uses are nickels. How many nickels did he use to buy the toy?

_____ nickels

HOT Problem Tamika has 14 nickels. How much money does she have? How many dimes are used to show that same amount?

Name ....

My Homework

Lesson 9
Count by Fives Using Nickels

Homework Helper

eHelp
Need help? connectED.mcgraw-hill.com

You can count by fives to count nickels.

5¢ 10¢ 15¢ 20¢ 25¢ 30¢ in all

Practice

Count by fives. Write the numbers.
Write how much there is in all.

1.

______¢ ______¢ ______¢ in all

2.

______¢ ______¢ ______¢ ______¢ ______¢ in all

Count by fives. Write the numbers. Write how much there is in all.

3.

______¢ ______¢ ______¢ ______¢ ______¢

______¢ ______¢ ______¢ ______¢ ______¢

in all

---

4. Jada buys a doll for 60¢. If she uses all nickels, how many nickels will she use?

______ nickels

Test Practice

5. Count the coins. How much money is shown?

- ○ 10¢
- ○ 15¢
- ○ 20¢
- ○ 25¢

Math at Home Give your child several nickels. Have him or her count the coins to tell you how much money they have in all.

Name

Use Models to Compare Numbers

Lesson 10

ESSENTIAL QUESTION
How can I use place value?

Explore and Explain

Watch

Tools

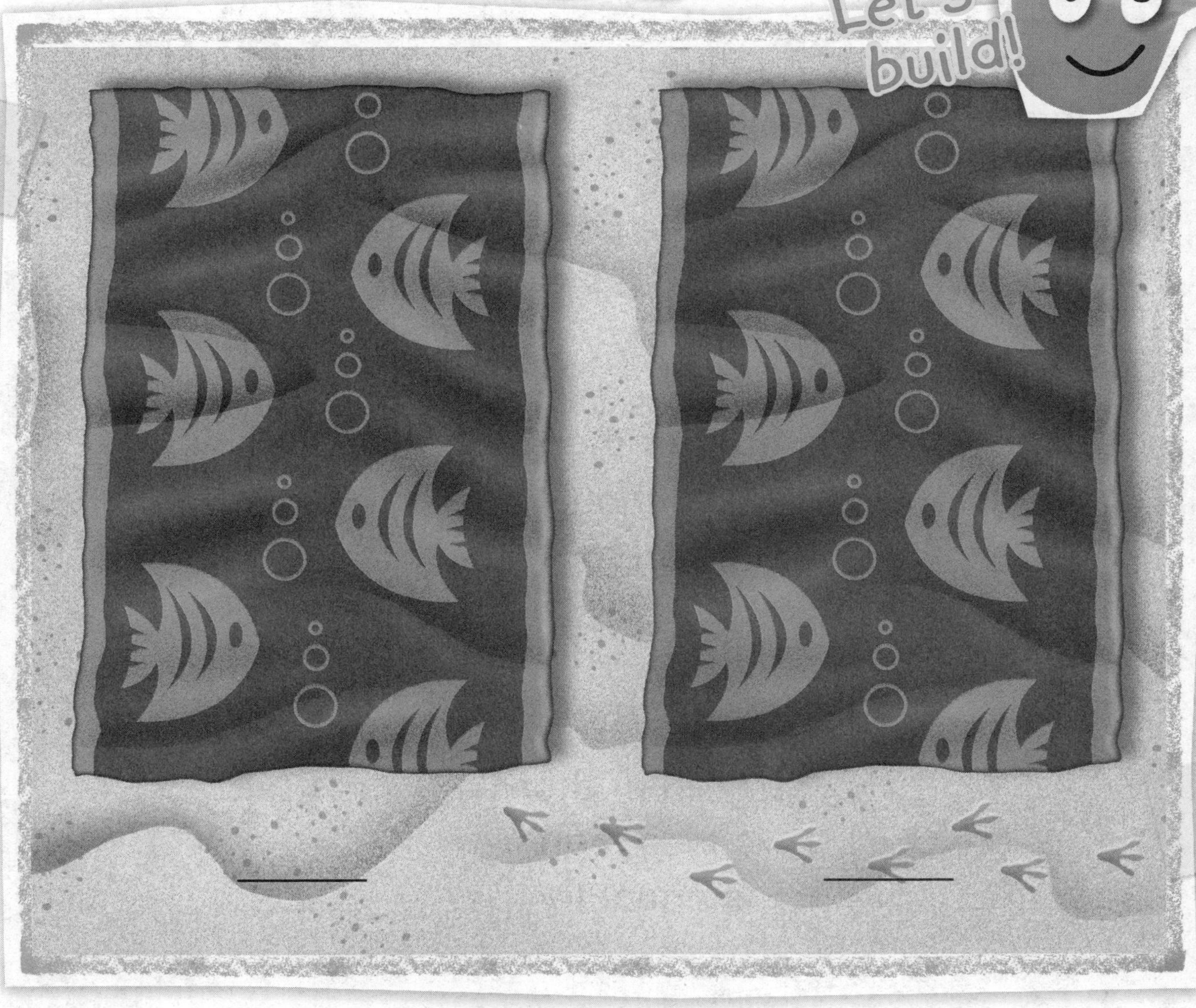

Teacher Directions: Use and . Show 38 on one towel. Show 19 on the other towel. Write the numbers. Circle the number that is greater. Place an X on the number that is less.

See and Show

Processes & Practices

Models can be used to compare numbers.

greater than	**less than**	**equal to**
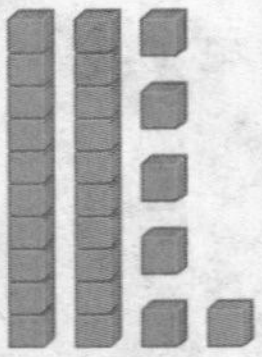 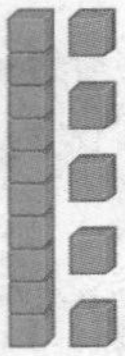	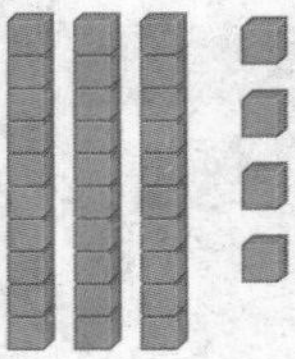 	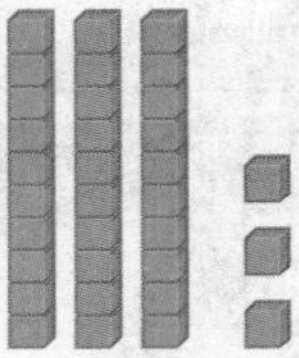 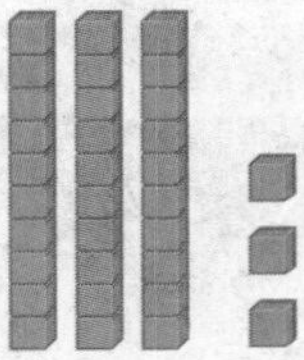
26 ___ 15 ___	34 ___ 35 ___	33 ___ 33 ___
26 is greater than 15.	34 is less than 35.	33 is equal to 33.

Use and . Write each number.
Circle *is greater than, is less than,* or *is equal to.*

1. 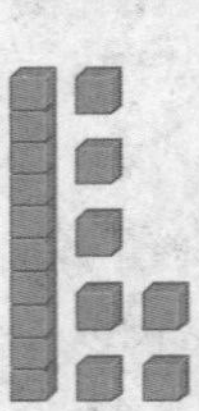

is greater than
is less than
is equal to

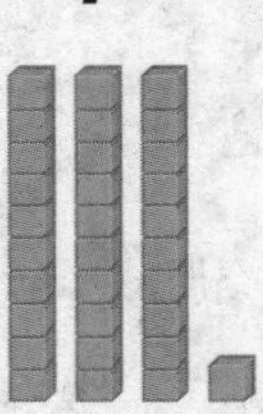

______ ______

2. 

is greater than
is less than
is equal to

______ ______

Talk Math How do you know 48 is greater than 38?

Name ....

On My Own

Use [tens rod] and [ones cube]. Write each number.
Circle *is greater than, is less than,* or *is equal to.*

3.

is greater than
is less than
is equal to

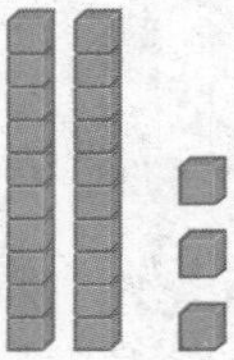

______ ______

4.

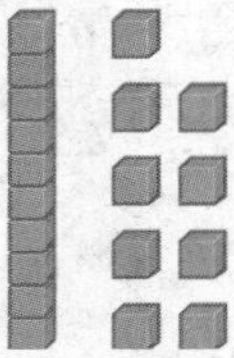

is greater than
is less than
is equal to

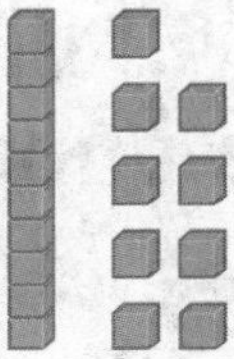

______ ______

5.

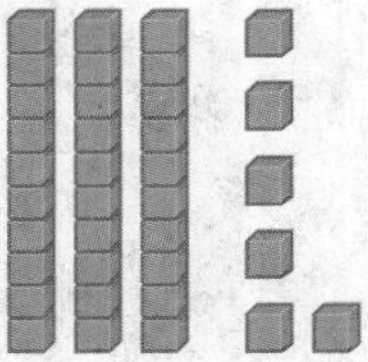

is greater than
is less than
is equal to

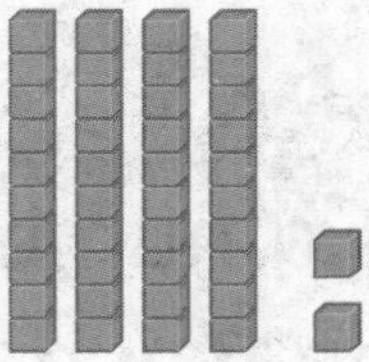

______ ______

Use [tens rod] and [ones cube]. Circle yes or no.

6. Is 41 less than 91?
yes no

7. Is 53 equal to 53?
yes no

8. Is 37 greater than 50?
yes no

9. Is 56 less than 65?
yes no

Processes & Practices

Problem Solving

10. I have fewer than 25 acorns.
I have more than 21 acorns.
Draw how many acorns I could have.

11. Beth's puppy is 63 days old.
Jon's puppy is 12 days old.
Whose puppy is older?

______________ puppy

HOT Problem Hayden says the model shows a number less than 34. Tell why Hayden is wrong. Make it right.

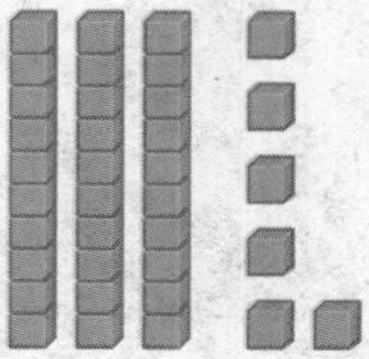

Name ________________

My Homework

Lesson 10

Use Models to Compare Numbers

Homework Helper

eHelp

Need help? connectED.mcgraw-hill.com

You can use models to compare numbers.

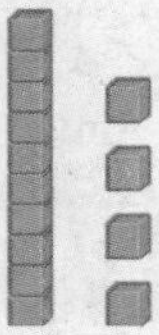

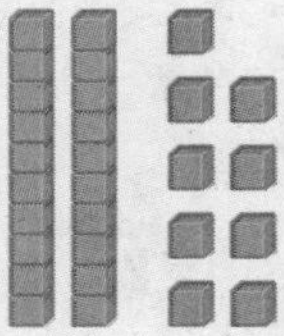

14 is greater than / (is less than) / is equal to 29

Practice

Write each number. Circle *is greater than*, *is less than*, or *is equal to*.

1\.

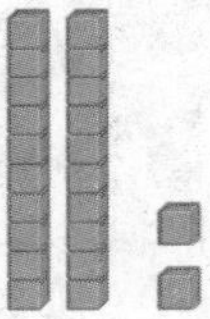

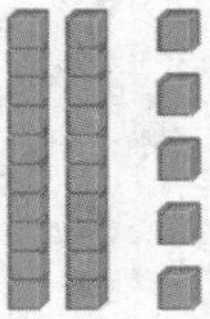

_______ is greater than / is less than / is equal to _______

2\.

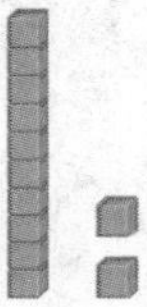

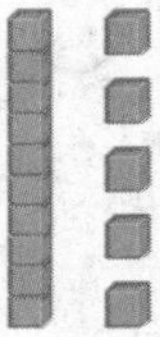

_______ is greater than / is less than / is equal to _______

Write each number. Circle *is greater than*, *is less than*, or *is equal to*.

3. 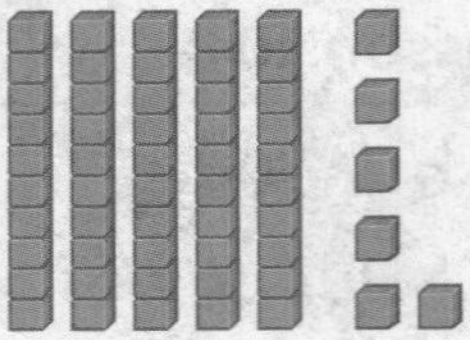

is greater than
is less than
is equal to

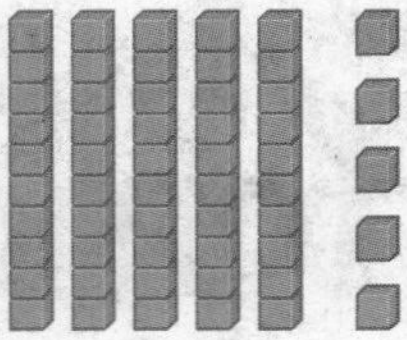

______ ______

4. 

is greater than
is less than
is equal to

______ ______

5. Aubrey read 29 books and Eli read 52 books. Who read less books?

____

Test Practice

6. A number is greater than 11 and less than 13. What is the number?

10 ○ 12 ○ 15 ○ 20 ○

Math at Home Write a number. Have your child name two numbers that are less than the number and two numbers that are greater. Ask your child what number is equal to the number.

Name ....

Use Symbols to Compare Numbers

Lesson 11

ESSENTIAL QUESTION
How can I use place value?

Explore and Explain

Watch Tools

Friday ________

Saturday ________

greater than (>) less than (<) equal to (=)

Teacher Directions: A store sold 32 toys on Friday. It sold 16 toys on Saturday. Use [ten rod] and [unit cube] to show each group. Write the numbers. Compare the numbers. Circle greater than (>), less than (<), or equal to (=).

See and Show

Processes & Practices

Symbols can be used to compare numbers.

greater than (>)

tens	ones
2	8
3	7

37 is greater than 28.

37 (>) 28

less than (<)

tens	ones
4	9
5	1

49 is less than 51.

49 (<) 51

equal to (=)

tens	ones
1	5
1	5

15 is equal to 15.

15 (=) 15

Compare. Write >, <, or =.

1.

tens	ones
6	3
3	5

63 ◯ 35

2.

tens	ones
2	1
2	1

21 ◯ 21

3.

tens	ones
1	2
2	4

12 ◯ 24

4.

tens	ones
3	5
4	3

35 ◯ 43

Talk Math Name a number that is greater than 38 and less than 46.

Name ....

Helpful Hint
To compare two numbers, think of the amount of tens and ones in each number.

On My Own

Compare. Write >, <, or =.

5.

tens	ones
6	5
5	4

65 ◯ 54

6.

tens	ones
3	3
3	3

33 ◯ 33

7. 45 ◯ 75

8. 80 ◯ 26

9. 12 ◯ 12

10. 95 ◯ 59

11. 85 ◯ 85

12. 60 ◯ 98

13. 62 ◯ 67

14. 49 ◯ 90

15. 96 ◯ 69

16. 42 ◯ 24

17. 53 ◯ 57

18. 41 ◯ 41

Circle the number less than the green number.

19. 80
84 79

20. 95
100 90

21. 23
9 99

Circle the number greater than the green number.

22. 65
61 71

23. 38
52 25

24. 6
10 0

Processes & Practices

Real World Problem Solving

25. There are 23 children riding on Brent's bus. Karie's bus has 32 children riding on it. Whose bus has more people riding on it?

____________ bus

26. Jon's kitten is 76 days old. Makala's kitten is 83 days old. Whose kitten is younger?

____________ kitten

Write Math What do these symbols mean: >, < and = ?

Name ....

My Homework

Lesson 11

Use Symbols to Compare Numbers

Homework Helper

eHelp **Need help?** connectED.mcgraw-hill.com

You can use symbols to compare numbers.

greater than (>)

tens	ones
2	1
1	8

21 is greater than 18.

21 > 18

less than (<)

tens	ones
5	9
6	0

59 is less than 60.

59 < 60

equal to =

tens	ones
2	4
2	4

24 is equal to 24.

24 = 24

Practice

Compare. Write >, <, or =.

1.

tens	ones
7	1
3	3

71 ◯ 33

2.

tens	ones
2	9
3	1

29 ◯ 31

3. 12 ◯ 14

4. 82 ◯ 82

Compare. Write >, <, or =.

5. 1 ◯ 50	6. 85 ◯ 62	7. 100 ◯ 100
8. 16 ◯ 8	9. 96 ◯ 62	10. 39 ◯ 42
11. 75 ◯ 55	12. 18 ◯ 96	13. 40 ◯ 37

14. Mike has 25 toys. Lucy has 31 toys.
Who has more toys?

____

Vocabulary Check

Circle the correct answer.

15. Which number is **equal to** (=) 88?

75 87 88

16. Which number is **less than** (<) 10?

15 9 11

17. Which number is **greater than** (>) 23?

25 16 22

Math at Home Write a number. Leave some space and write another number. Have your child place the correct symbol, >, <, or =, to compare the two numbers. Repeat several times with different numbers.

Name ....

Check My Progress

Vocabulary Check

Vocab

Write the correct word and symbol.

greater than (>) **less than (<)** **equal to (=)**

1. 16 is ______________________ 33.
2. 34 is ______________________ 34.
3. 72 is ______________________ 70.

Concept Check

Write the tens and ones. Then write the number.

4.

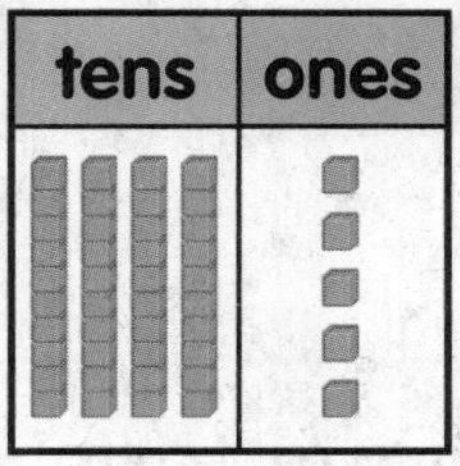

tens	ones

_____ forty-five

Write the number that is ten more.

5. 18, _____ 6. 79, _____ 7. 11, _____

Write the number that is ten less.

8. _____, 54 9. _____, 98 10. _____, 40

Compare. Write >, <, or =.

11.

tens	ones
2	2
2	3

22 ◯ 23

12.

tens	ones
5	9
5	9

59 ◯ 59

13. 16 ◯ 17

14. 18 ◯ 10

15. 21 ◯ 21

16. 42 ◯ 34

17. 59 ◯ 79

18. 96 ◯ 98

19. Armondo has 19 baseballs. He has 10 more jacks than baseballs. How many jacks does Armondo have?

_____ jacks

Test Practice

20. The boys have 65 toy train cars. How many tens and ones do they have?

60 tens 5 ones ◯

6 tens 5 ones ◯

6 tens 0 ones ◯

5 tens 6 ones ◯

Name

Numbers to 120

Lesson 12

ESSENTIAL QUESTION

How can I use place value?

Explore and Explain

Watch Tools

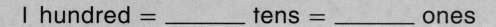

I hundred = ______ tens = ______ ones

Teacher Directions: Use [hundred flat], [ten rod], and [one cube] to show 100 in three different ways. Write how many tens and how many ones you used to show 100.

See and Show

Processes & Practices

You can use hundreds, tens, and ones to show a number.

hundreds	tens	ones
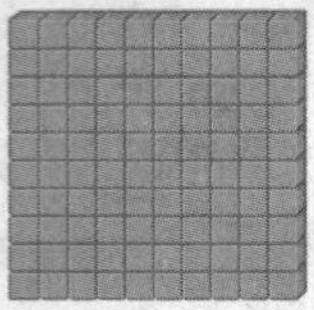		

hundreds	tens	ones
1	1	6

1 hundred 1 ten 6 ones

116

one hundred sixteen

Use Work Mat 8, , , and , to show each number. Then write the number two ways.

1.

hundreds	tens	ones

hundreds	tens	ones

______

one hundred fourteen

Talk Math How many hundreds, tens, and ones are in the number 102?

Name ....

On My Own

Use Work Mat 8, , , and , to show each number. Then write the number two ways.

2.

hundreds	tens	ones
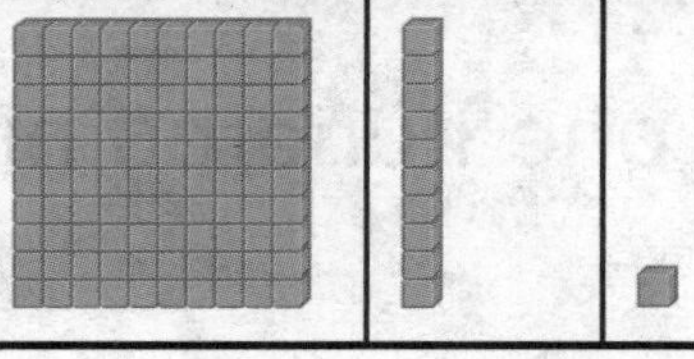		

hundreds	tens	ones

______

one hundred eleven

3.

hundreds	tens	ones
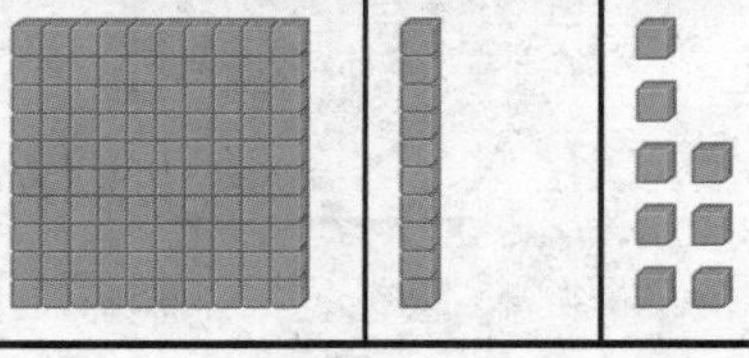		

hundreds	tens	ones

______

one hundred eighteen

4.

hundreds	tens	ones
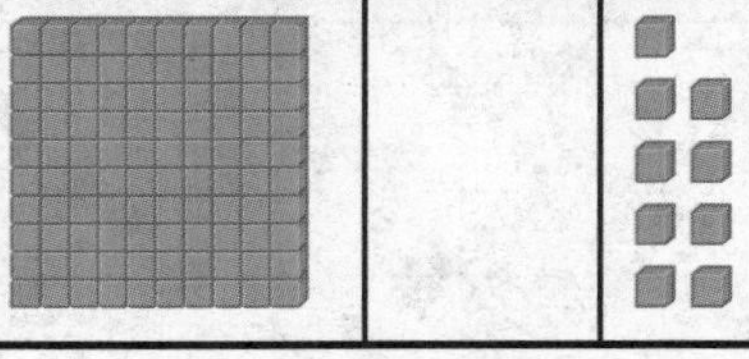		

hundreds	tens	ones

______

one hundred nine

Problem Solving

Processes & Practices

5. Jessica models a number in the chart.
What number does she model?

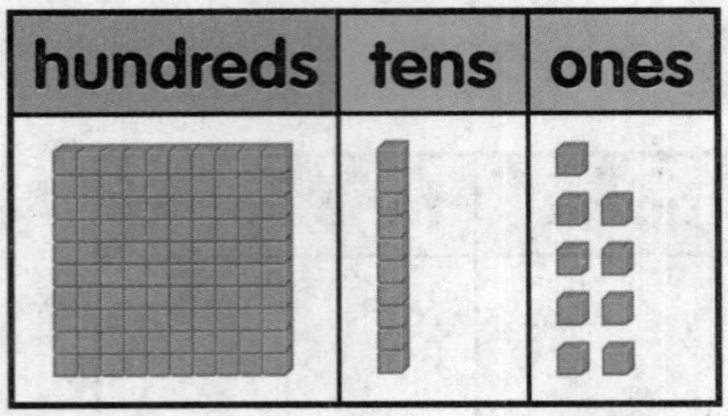

_____ one hundred nineteen

6. Evie collects stickers. She has 1 hundred, 1 ten, and 2 ones.
How many stickers does Evie have?

_____ stickers

Write Math Explain the difference between the numbers 15 and 115.

Name ....

My Homework

Lesson 12
Numbers to 120

Homework Helper

eHelp Need help? connectED.mcgraw-hill.com

You can use hundreds, tens, and ones to show a number.

hundreds	tens	ones
		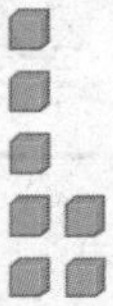

hundreds	tens	ones
1	1	7

1 hundred 1 ten 7 ones

117

one hundred seventeen

Practice

Write the number two ways.

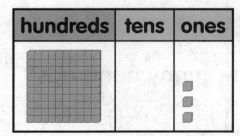

1.

hundreds	tens	ones
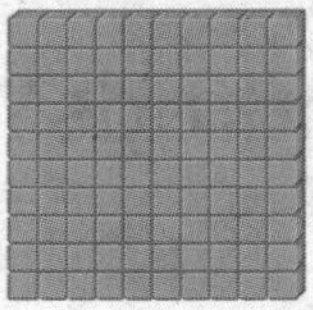		

hundreds	tens	ones

____

one hundred three

Write the number two ways.

2.

hundreds	tens	ones

hundreds	tens	ones

____

one hundred thirteen

3.

hundreds	tens	ones

hundreds	tens	ones

____

one hundred

4. Owen has 1 hundred, 0 tens, and 1 one marbles. How many marbles does Owen have?

_______ marbles

Vocabulary Check

Circle the correct answer.

5. Which number shows how many hundreds?

1 1 0

Math at Home Write the number 104. Ask your child to tell you how many hundreds, tens, and ones.

Name

Count to 120

Lesson 13

ESSENTIAL QUESTION
How can I use place value?

Explore and Explain

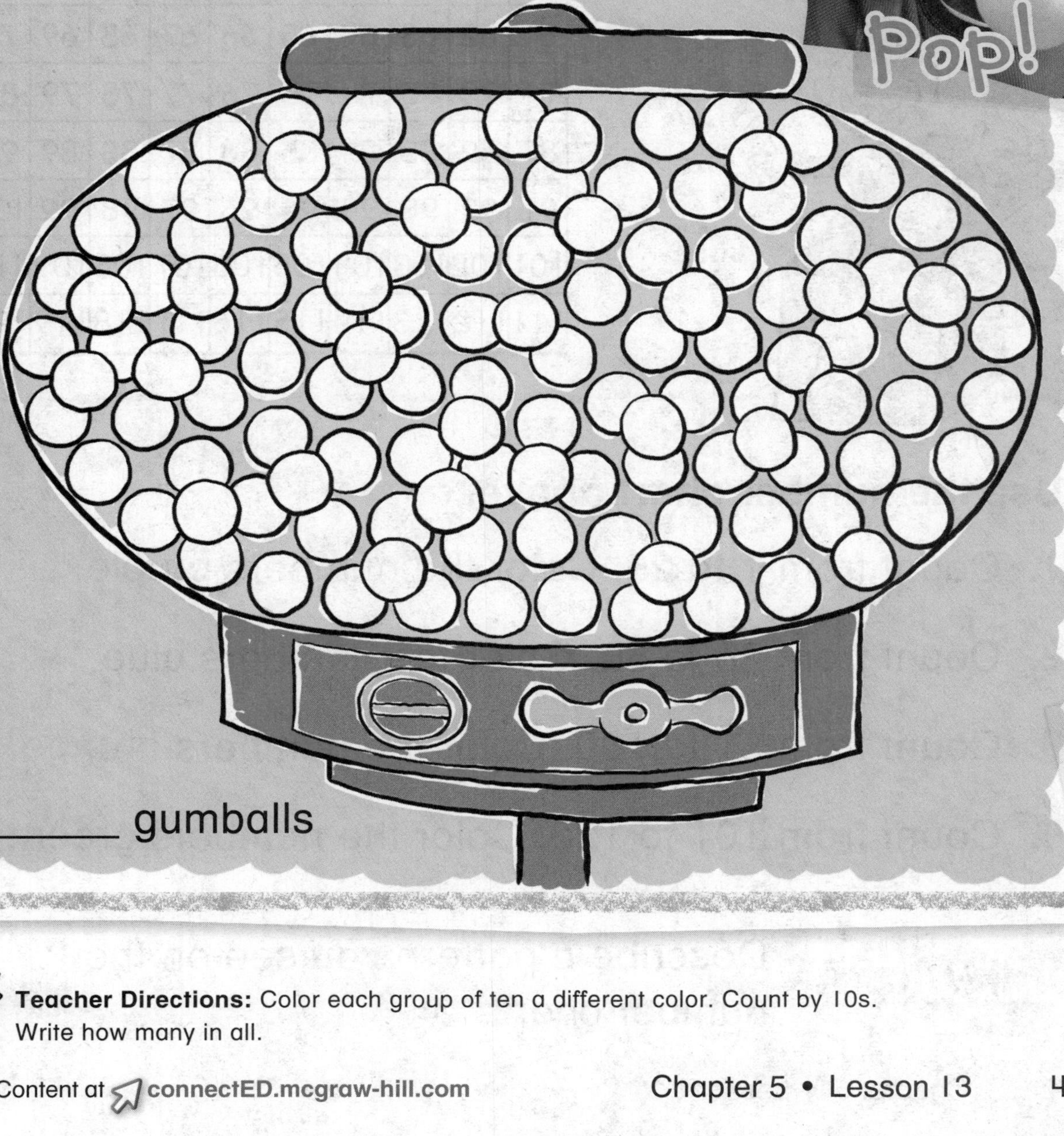

_______ gumballs

Teacher Directions: Color each group of ten a different color. Count by 10s. Write how many in all.

See and Show

Processes & Practices

You can count numbers in order to 120.

Count from 21 to 28.
Circle the number that comes next.

27 (29) 30

1	2	3	4	5	6	7	8	9	10
11	12	13	14	15	16	17	18	19	20
21	22	23	24	25	26	27	28	29	30
31	32	33	34	35	36	37	38	39	40
41	42	43	44	45	46	47	48	49	50
51	52	53	54	55	56	57	58	59	60
61	62	63	64	65	66	67	68	69	70
71	72	73	74	75	76	77	78	79	80
81	82	83	84	85	86	87	88	89	90
91	92	93	94	95	96	97	98	99	100
101	102	103	104	105	106	107	108	109	110
111	112	113	114	115	116	117	118	119	120

Use the number chart above.

1. Count from 1 to 25. Color the numbers purple.
2. Count from 26 to 50. Color the numbers blue.
3. Count from 51 to 100. Color the numbers orange.
4. Count from 101 to 120. Color the numbers green.

Talk Math Describe a pattern you see on the number chart.

Name ....

On My Own

Use the number chart. Circle the number that comes next.

1	2	3	4	5	6	7	8	9	10
11	12	13	14	15	16	17	18	19	20
21	22	23	24	25	26	27	28	29	30
31	32	33	34	35	36	37	38	39	40
41	42	43	44	45	46	47	48	49	50
51	52	53	54	55	56	57	58	59	60
61	62	63	64	65	66	67	68	69	70
71	72	73	74	75	76	77	78	79	80
81	82	83	84	85	86	87	88	89	90
91	92	93	94	95	96	97	98	99	100
101	102	103	104	105	106	107	108	109	110
111	112	113	114	115	116	117	118	119	120

5. Count from 1 to 13.
What number comes next?

10 14 20

6. Count from 16 to 23.
What number comes next?

24 23 22

7. Count from 29 to 35.
What number comes next?

30 33 36

8. Count from 65 to 77.
What number comes next?

11 78 79

Circle the group that shows the numbers in counting order.

9. 52, 36, 48, 65 97, 98, 99, 100

10. 84, 85, 86, 87 3, 30, 13, 31

Problem Solving

Processes & Practices

11. This number is before 57. It is after 53. It has 5 tens and 5 more. What number is it?

______

12. This number is before 76. It is after 72. It has 7 tens and 4 ones. What number is it?

______

Write Math What are some things you see on a number chart that are alike?

Name ....

My Homework

Lesson 13

Count to 120

Homework Helper

eHelp Need help? connectED.mcgraw-hill.com

Count from 11 to 22. What number comes next?

22 (23) 24

Count from 73 to 85. What number comes next?

(86) 87 88

1	2	3	4	5	6	7	8	9	10
11	12	13	14	15	16	17	18	19	20
21	22	23	24	25	26	27	28	29	30
31	32	33	34	35	36	37	38	39	40
41	42	43	44	45	46	47	48	49	50
51	52	53	54	55	56	57	58	59	60
61	62	63	64	65	66	67	68	69	70
71	72	73	74	75	76	77	78	79	80
81	82	83	84	85	86	87	88	89	90
91	92	93	94	95	96	97	98	99	100
101	102	103	104	105	106	107	108	109	110
111	112	113	114	115	116	117	118	119	120

Practice

Use the number chart. Circle the number that comes next.

1. Count from 78 to 90. What number comes next?

   80 93 91

2. Count from 112 to 118. What number comes next?

   117 108 119

Use the number chart.

3. Count from 1 to 33. Color the numbers purple.

4. Count from 34 to 66. Color the numbers orange.

5. Count from 67 to 99. Color the numbers green.

6. Count from 100 to 119. Color the numbers blue.

1	2	3	4	5	6	7	8	9	10
11	12	13	14	15	16	17	18	19	20
21	22	23	24	25	26	27	28	29	30
31	32	33	34	35	36	37	38	39	40
41	42	43	44	45	46	47	48	49	50
51	52	53	54	55	56	57	58	59	60
61	62	63	64	65	66	67	68	69	70
71	72	73	74	75	76	77	78	79	80
81	82	83	84	85	86	87	88	89	90
91	92	93	94	95	96	97	98	99	100
101	102	103	104	105	106	107	108	109	110
111	112	113	114	115	116	117	118	119	120

Test Practice

7. Ally started counting at 47. Which shows the next 4 numbers she counted?

- ○ 57, 67, 78, 79
- ○ 48, 49, 50, 51
- ○ 82, 81, 83, 90
- ○ 48, 50, 52, 54

Math at Home Have your child choose a number on the hundred chart and show you how to count to 120 from that number.

Name ....

Read and Write Numbers to 120

Lesson 14

ESSENTIAL QUESTION
How can I use place value?

Explore and Explain

1	2	3	4	5	6	7	8	9	10
11	12	13	14	15	16	17	18	19	20
	22	23	24	25	26	27	28	29	30
31	32	33	34	35	36	37	38	39	40
41	42	43	44	45	46	47	48	49	50
51	52	53	54		56	57	58	59	60
61	62	63	64	65	66	67	68	69	70
71	72	73	74	75	76	77	78	79	80
81	82	83	84	85	86	87	88	89	90
91	92	93	94	95	96	97	98	99	100
101	102	103	104	105	106	107		109	110
111	112	113	114	115	116	117	118	119	120

Teacher Directions: There are some numbers missing from the chart. Find and write these missing numbers. Read the numbers from 1 to 120. Now read the numbers from 71 to 80. Color those numbers yellow.

See and Show

Processes & Practices

You can read and write numbers to 120.

1	2	3	4	5	6	7	8	9	10
11	12	13	14	15	16	17	18	19	20
21	22	23	24	25	26	27	28	29	30
31	32	33	34	35	36	37	38	39	40
41	42	43	44	45	46	47	48	49	50
51	52	53	54	55	56	57	58	59	60
61	62	63	64	65	66	67	68	69	70
71	72	73	74	75	76	77	78	79	80
81	82	83	84	85	86	87	88	89	90
91	92	93	94	95	96	97	98	99	100
101	102	103	104	105	106	107	108	109	110
111	112	113	114		116	117	118	119	120

Helpful Hint
Find 101 on the number chart.

Write the missing number. Read the numbers from 101 to 120. Color those numbers green.

115 ______

Write the missing numbers. Read the numbers.

1. 103, ______, 105, 106, 107, ______, 109

2. 114, 115, 116, ______, 118, 119, ______

3. 97, ______, 99, ______, ______, 102, 103

4. Write fifty-nine using numbers. ______

Talk Math Tell how you read 116. Explain how you write one hundred ten in numbers.

Name

On My Own

Write the missing numbers. Read the numbers.

5.

1	2	3		5	6	7	8	9	10
11		13	14	15	16		18	19	20
21	22	23	24		26	27	28	29	
31	32		34	35		37	38	39	40
	42	43	44	45	46	47		49	50
51	52		54		56	57	58		60
61		63	64	65	66		68	69	70
	72	73	74	75	76	77	78	79	80
81	82	83		85	86	87	88	89	
91	92	93	94	95		97		99	100
101		103	104		106	107		109	110
	112		114	115	116	117		119	

Write the number.

6. _______ eighty-three

7. _______ thirty-seven

8. _______ one hundred nineteen

9. _______ one hundred five

Problem Solving

Processes & Practices

10. A number was erased. What number is missing?

111 112 113 ___ 115 116 117

______

11. Graelyn is thinking of a number. It comes before 88 and after 86. What is the number?

______

HOT Problem Molly read the numbers to the right as sixty-five, fifty-seven, and eighty-five. Tell why Molly is wrong. Make it right.

65, 75, 85

Name ....

My Homework

Lesson 14

Read and Write Numbers to 120

Homework Helper

eHelp Need help? connectED.mcgraw-hill.com

1	2	3	4	5	6	7	8	9	10
11	12	13	14	15	16	17	18	19	20
21	22	23	24	25	26	27	28	29	30
31	32	33	34	35	36	37	38	39	40
41	42	43	44	45	46	47	48	49	50
51	52	53	54	55	56	57	58	59	60
61	62	63	64	65	66	67	68	69	70
71	72	73	74	75	76	77	78	79	80
81	82	83	84	85	86	87	88	89	90
91	92	93	94	95	96	97	98	99	100
101	102	103	104	105	106	107	108	109	110
111	112	113	114	115	116	117	118	119	120

You can read and write numbers to 120.

The numbers from 88 to 93 are written as 88, 89, 90, 91, 92, and 93.

The number one hundred six is written as 106.

Write the missing numbers. Read the numbers.

1. 85, ______, 87, 88, ______, 90

2. ______, 101, 102, ______, ______

3. 71, 72, ______, 74, 75, ______, 77

Read and write the missing numbers.

4. 114, ______, 116, 117, 118, ______, 120

Write the number.

5. ______ twenty-seven

6. ______ thirteen

7. ______ ninety

8. ______ seventy-four

9. ______ sixty-two

10. ______ one hundred eleven

11. Matt is thinking of a number. It comes before 39 and after 37. What is the number?

______

Test Practice

12. Freddie counted to 103. What number comes next?

94	102	104	114
○	○	○	○

Math at Home Write a number from 1 to 120. Have your child read the number to you. Then have your child count the next ten numbers from your number.

Name ______________________________

My Review

Chapter 5
Place Value

Vocabulary Check

Vocab

Draw a line to the correct word.

1. tens	=
2. equal to	<
3. hundred	>
4. less than	10 ones
5. regroup	10 tens
6. greater than	19 ones = 1 ten and 9 ones

Concept Check

Write the number different ways.

7.

tens	ones
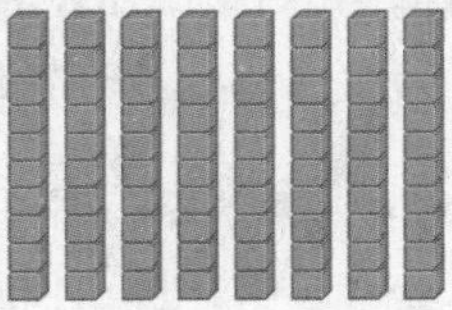	

tens	ones

______

eighty-three

Write the number different ways.

8.

hundreds	tens	ones

hundreds	tens	ones

____

one hundred five

Use the number chart. Circle the number that comes next.

9. Count from 34 to 47.
What number comes next?

46 47 48

1	2	3	4	5	6	7	8	9	10
11	12	13	14	15	16	17	18	19	20
21	22	23	24	25	26	27	28	29	30
31	32	33	34	35	36	37	38	39	40
41	42	43	44	45	46	47	48	49	50
51	52	53	54	55	56	57	58	59	60
61	62	63	64	65	66	67	68	69	70
71	72	73	74	75	76	77	78	79	80
81	82	83	84	85	86	87	88	89	90
91	92	93	94	95	96	97	98	99	100
101	102	103	104	105	106	107	108	109	110
111	112	113	114	115	116	117	118	119	120

10. Count from 86 to 99.
What number comes next?

98 100 90

11. Circle the group that shows these numbers in counting order.

58, 59, 60, 61 60, 58, 61, 59

12. Liz has 26 raisins in her lunch.
How many tens and ones are in 26?

_____ tens and _____ ones

Name ______________________

Problem Solving

13. Brandon has 5 baskets. Each basket has 10 peanuts. How many peanuts are there in all?

_______ peanuts

14. Tommy buys 10 marbles each week. How many marbles will he have in 6 weeks?

_______ marbles

15. Allison makes 3 groups of 10 muffins. She also makes 7 more muffins. How many muffins does she make in all?

_______ muffins

Test Practice

16. Read the numbers. What number is missing?

96, 97, _____, 99, 100, 101

95	98	102	104
○	○	○	○

Reflect

Chapter 5
Answering the Essential Question

Show what you know about place value.

Write the number.

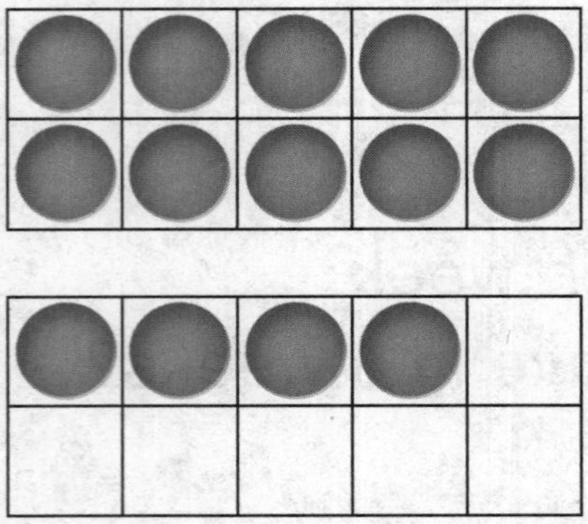

______
fourteen

Compare.
Use >, <, or =.

19 ◯ 19

86 ◯ 23

4 ◯ 14

ESSENTIAL QUESTION
How can I use place value?

How many tens and ones?

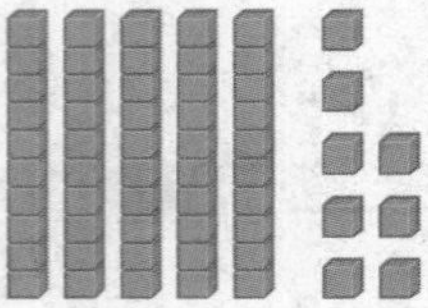

____ tens ____ ones

______
fifty-eight

How many hundreds, tens, and ones?

hundreds	tens	ones
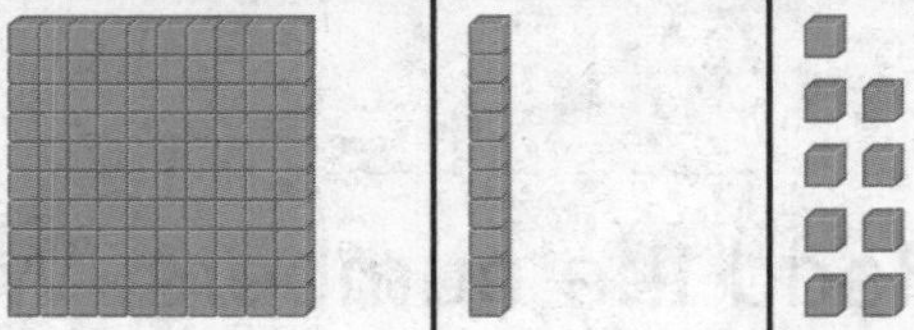		
____ hundred	____ ten	____ ones

Chapter

6 Two-Digit Addition and Subtraction

ESSENTIAL QUESTION

How can I add and subtract two-digit numbers?

My Favorite Activities!

Watch a video!

Name

Chapter 6 Project

Add Two-Digit Numbers

1. Write the two-digit addition concept on the line below.
2. Write a number sentence showing an example of the two-digit addition concept.
3. Solve the number sentence by writing the sum.
4. Write all of the steps used in solving the addition problem.

Name ....

Am I Ready?

Write how many tens and ones.

1. 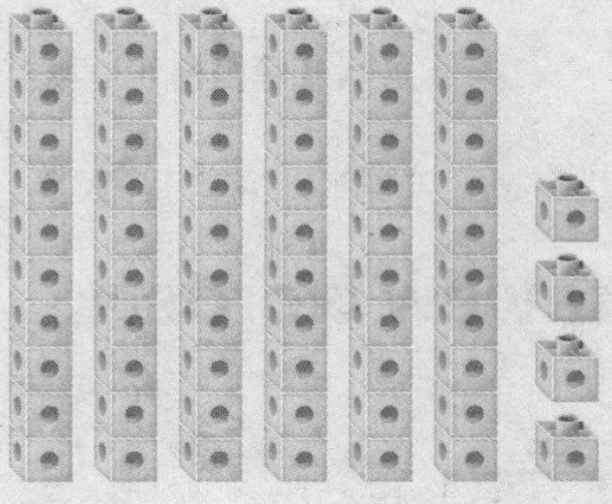

____ tens ____ ones =

____

2. 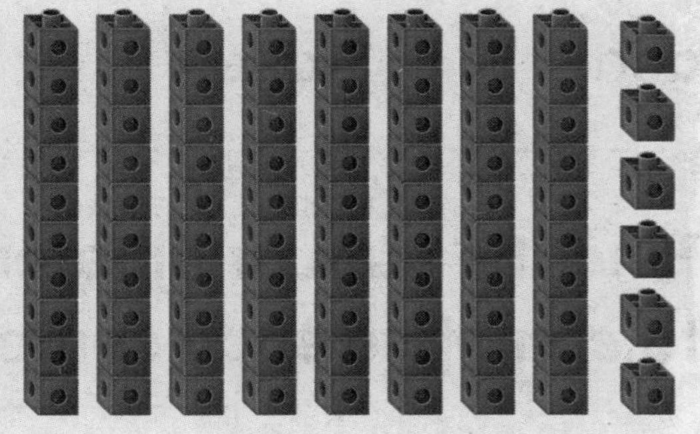

____ tens ____ ones =

____

Add.

3. 9 + 1 = ______

4. 4 + 2 = ______

Subtract.

5. 6 − 2 = ______

6. 7 − 6 = ______

7. Maggie has 18 stickers. She gives 9 of them away. How many stickers does Maggie have left?

______ stickers

How Did I Do?

Shade the boxes to show the problems you answered correctly.

1	2	3	4	5	6	7

Name ....

My Math Words

Vocab

Review Vocabulary

add ones subtract tens

Find the sum or difference. Use the review words to complete each sentence.

26	19
+ 2	− 7

The number 28 is 2 ____________ and 8 ones.

I can ____________ to find the sum.

The number 12 is 1 ten and 2 ____________.

I can ____________ to find the difference.

My Vocabulary Cards

Vocab

abc

Processes & Practices

Teacher Directions:
Ideas for Use

- Ask students to use the blank cards to write their own vocabulary cards.
- Have students use blank cards to write basic addition and subtraction facts. Have students write the answer on the back of each card.

My Foldable

FOLDABLES® Follow the steps on the back to make your Foldable.

FOLDABLES® Study Organizer

1 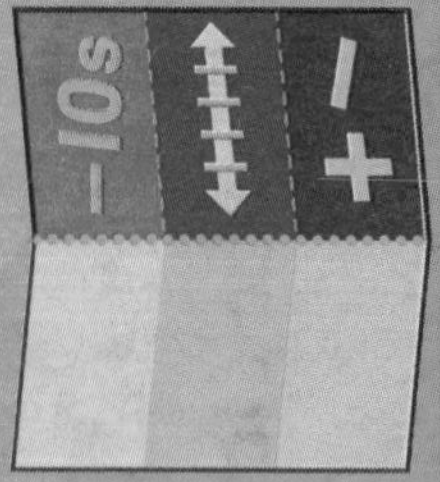

2 

3 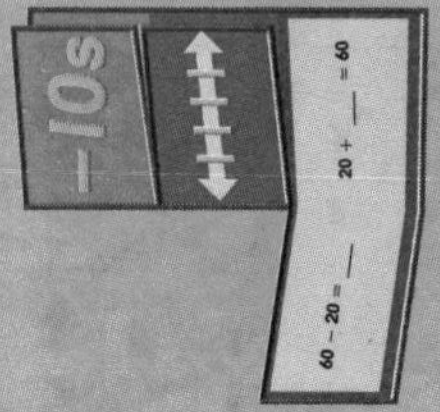

6 tens − 2 tens = ___ tens

60 − 20 = ___

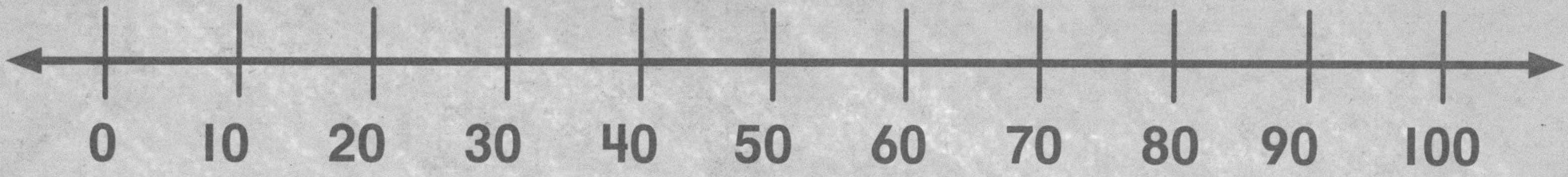

60 − 20 = ___

60 − 20 = ___

20 + ___ = 60

Name ....

Add Tens

Lesson 1

ESSENTIAL QUESTION
How can I add and subtract two-digit numbers?

Explore and Explain

Watch Tools

3 tens + 2 tens = ____ tens

30 + 20 = ____

Teacher Directions: Use [ten-rod] to model. A family buys 30 pieces of firewood. They then buy 20 more pieces. How many pieces of firewood did they buy in all? Draw the rods you used. Write how many tens. Explain to a classmate how you found the answer.

See and Show

Processes & Practices

Helpful Hint
To find 20 + 20, add 2 tens + 2 tens which equals 4 tens or 40.

Find 20 + 20 by adding the tens.

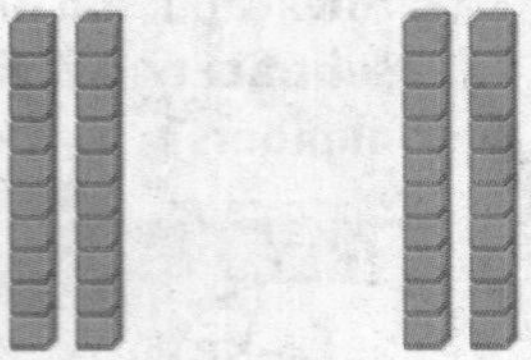

2 tens + 2 tens = __4__ tens

20 + 20 = __40__

Add. Use  to help.

1. 4 tens + 2 tens = ______ tens   40 + 20 = ______

2. 6 tens + 1 ten = ______ tens   60 + 10 = ______

3. 5 tens + 2 tens = ______ tens   50 + 20 = ______

4. 7 tens + 2 tens = ______ tens   70 + 20 = ______

5. 8 tens + 1 ten = ______ tens   80 + 10 = ______

Talk Math How does knowing 2 + 5 help you find 20 + 50?

Name ....

On My Own

Add. Use  to help.

6. 7 tens + 1 ten = ______ tens 70 + 10 = ______

7. 2 tens + 3 tens = ______ tens 20 + 30 = ______

8. 1 ten + 3 tens = ______ tens 10 + 30 = ______

9. 5 tens + 4 tens = ______ tens 50 + 40 = ______

10. 6 tens + 1 ten = ______ tens 60 + 10 = ______

11. 5 tens + 3 tens = ______ tens 50 + 30 = ______

12.

$$\begin{array}{r} 3 \text{ tens} \\ + \ 4 \text{ tens} \\ \hline \text{tens} \end{array} \qquad \begin{array}{r} 30 \\ + \ 40 \\ \hline \end{array}$$

13.

$$\begin{array}{r} 1 \text{ ten} \\ + \ 5 \text{ tens} \\ \hline \text{tens} \end{array} \qquad \begin{array}{r} 10 \\ + \ 50 \\ \hline \end{array}$$

14.

$$\begin{array}{r} 1 \text{ ten} \\ + \ 2 \text{ tens} \\ \hline \text{tens} \end{array} \qquad \begin{array}{r} 10 \\ + \ 20 \\ \hline \end{array}$$

15.

$$\begin{array}{r} 2 \text{ tens} \\ + \ 6 \text{ tens} \\ \hline \text{tens} \end{array} \qquad \begin{array}{r} 20 \\ + \ 60 \\ \hline \end{array}$$

Problem Solving

Processes & Practices

16. There are 50 boys and 20 girls at basketball camp. How many children are at basketball camp in all?

_____ children

17. There are 30 children at dance class on Tuesday. 30 other children are at dance class on Wednesday. How many children are at dance class in all?

_____ children

Write Math Will there be any ones in the answer when you add 50 + 40? Explain.

Name ______________________

My Homework

Lesson 1
Add Tens

Homework Helper

eHelp

Need help? connectED.mcgraw-hill.com

Find 60 + 20.

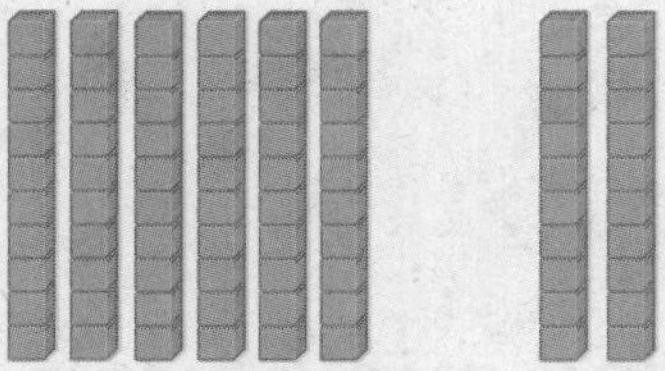

Helpful Hint
To find 60 + 20, add 6 tens and 2 tens.

6 tens + 2 tens = 8 tens

60 + 20 = 80

Practice

Add.

1. 3 tens + 4 tens = ______ tens   30 + 40 = ______

2. 1 ten + 2 tens = ______ tens   10 + 20 = ______

3. 5 tens + 4 tens = ______ tens   50 + 40 = ______

4. 3 tens + 3 tens = ______ tens   30 + 30 = ______

5. 3 tens + 5 tens = ______ tens   30 + 50 = ______

Add.

6. 2 tens + 2 tens = ___ tens

$$\begin{array}{r} 20 \\ +\ 20 \\ \hline \end{array}$$

7. 7 tens + 1 ten = ___ tens

$$\begin{array}{r} 70 \\ +\ 10 \\ \hline \end{array}$$

8. 2 tens + 3 tens = ___ tens

$$\begin{array}{r} 20 \\ +\ 30 \\ \hline \end{array}$$

9. 3 tens + 6 tens = ___ tens

$$\begin{array}{r} 30 \\ +\ 60 \\ \hline \end{array}$$

10. Today 40 people played miniature golf in the afternoon. 30 more people played it in the evening. How many people played miniature golf in all?

___ people

Test Practice

11. 4 tens + 4 tens = ___

- ○ 6 tens
- ○ 7 tens
- ○ 8 tens
- ○ 9 tens

Math at Home Have your child tell you how knowing 4 + 2 helps him or her find 40 + 20.

Name ....

Count On Tens and Ones

Lesson 2

ESSENTIAL QUESTION
How can I add and subtract two-digit numbers?

Explore and Explain

Watch Tools

$35 + 3 = \underline{38}$

Teacher Directions: Use [ten-rod] and [unit] to model. There are 35 people at a baseball game. 3 more people come to the game. How many people are at the game in all? Draw the rods and units that you used. Explain to a classmate how you found the answer.

See and Show

Find 26 + 3.
Count on by ones.

Start at 26.
Count 27, 28, 29.
The sum is 29.

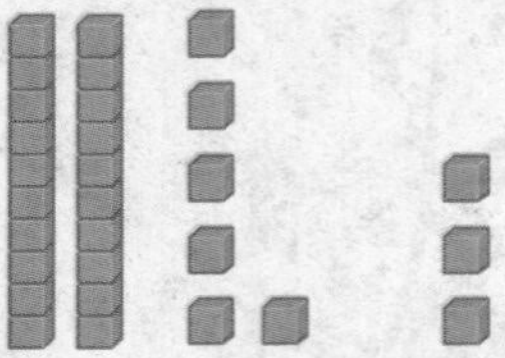

26 + 3 = 29

Find 26 + 30.
Count on by tens.

Start at 26.
Count 36, 46, 56.
The sum is 56.

26 + 30 = 56

Take me out to the ball game!

Count on to add. Use [ten rod] and [unit cube]. Write the sum.

1. 

47 + 2 = ______

2. 

47 + 20 = ______

3. 

13 + 3 = ______

4. 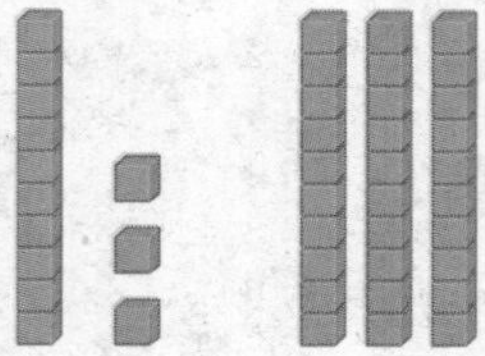

13 + 30 = ______

Talk Math How many tens do you count on to add 32 + 40? Explain.

Name ______________________________

On My Own

Count on to add. Use  and .
Write the sum.

5. 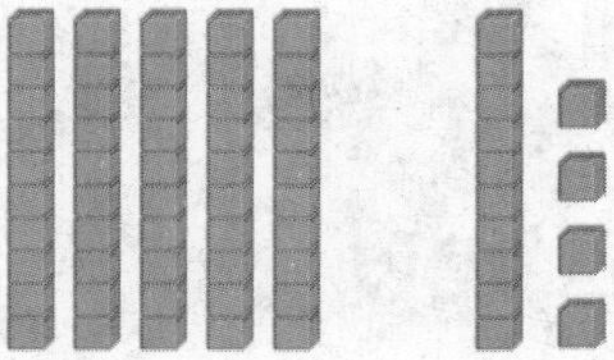

$50 + 14 =$ ______

6. 

$25 + 3 =$ ______

7. $30 + 22 =$ ______

8. $66 + 2 =$ ______

9. $53 + 20 =$ ______

10. $14 + 3 =$ ______

11. $51 + 3 =$ ______

12. $20 + 76 =$ ______

13. $\begin{array}{r} 44 \\ +\ \ 3 \\ \hline \end{array}$

14. $\begin{array}{r} 10 \\ +\ 88 \\ \hline \end{array}$

15. $\begin{array}{r} 88 \\ +\ \ 1 \\ \hline \end{array}$

16. $\begin{array}{r} 33 \\ +\ \ 3 \\ \hline \end{array}$

17. $\begin{array}{r} 12 \\ +\ \ 2 \\ \hline \end{array}$

18. $\begin{array}{r} 79 \\ +\ 20 \\ \hline \end{array}$

Problem Solving

Processes & Practices

19. Percy sees 3 ants in his ant farm. He sees 12 more ants. How many ants does he see in all?

_____ ants

20. Kevin's team and Alonso's team each have 25 points. Kevin's team scores 3 more points.

Kevin's team has _____ points.

Alonso's team scores 30 more points.

Alonso's team has _____ points.

Write Math Find 54 + 3. Explain how you add the ones.

Name ....

My Homework

Lesson 2
Count On Tens and Ones

Homework Helper

eHelp Need help? connectED.mcgraw-hill.com

Find 42 + 2.
Count on by ones.

Start at 42.
Count 43, 44.
The sum is 44.

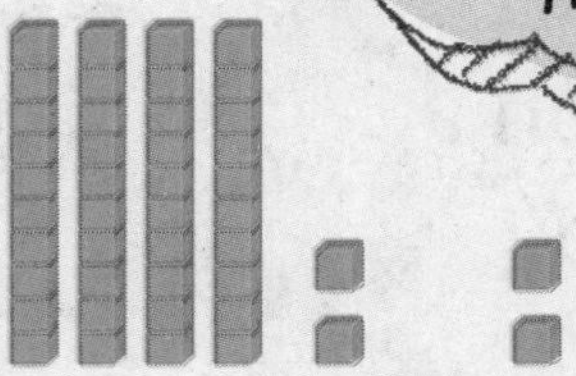

42 + 2 = 44

Find 30 + 15.
Count on by tens.

Start at 15.
Count 25, 35, 45.

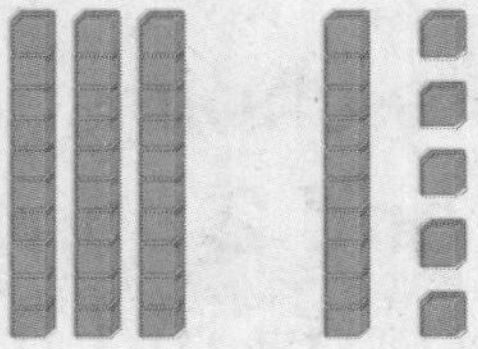

30 + 15 = 45

Practice

Count on to add. Write the sum.

1.

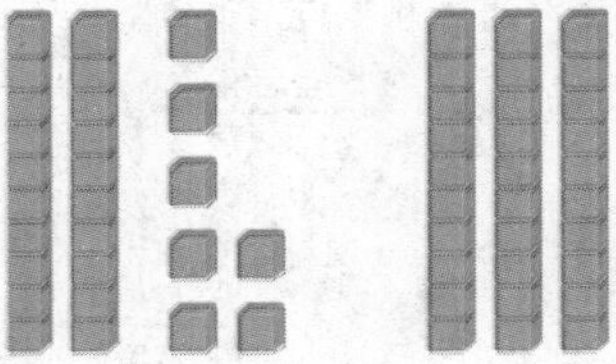

27 + 30 = ______

2.

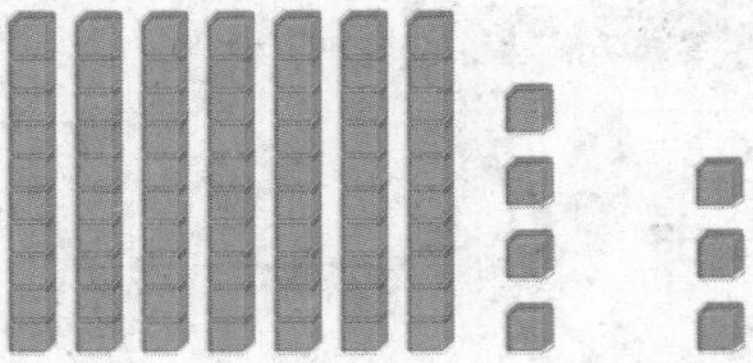

74 + 3 = ______

3. 66 + 3 = ______

4. 12 + 70 = ______

Count on to add. Write the sum.

5. 51 + 3 = ______

6. 87 + 1 = ______

7. 32 + 20	8. 46 + 10	9. 97 + 2
10. 26 + 10	11. 40 + 2	12. 64 + 20

13. The children's choir sang 17 songs on Wednesday and 2 songs on Thursday. How many songs did they sing in all?

______ songs

Test Practice

14. 70 + 20 = ______

78 ○ 85 ○ 88 ○ 90 ○

Math at Home Say a number between 10 and 50. Ask your child to count on by ones. Repeat. Ask your child to count on by tens.

Name ______________________

Add Tens and Ones

Lesson 3

ESSENTIAL QUESTION
How can I add and subtract two-digit numbers?

Explore and Explain

Watch Tools

tens	ones

_____ + _____ = _____

Teacher Directions: Use [rod] and [unit] to model. There are 24 children in a swimming pool. 3 more children get in. How many children are in the pool in all? Draw the rods and units you used. Write the numbers. Explain to a classmate how you found the answer.

See and Show

Processes & Practices

To find 25 + 2, add the ones. Then add the tens.

Step 1
Show each number.

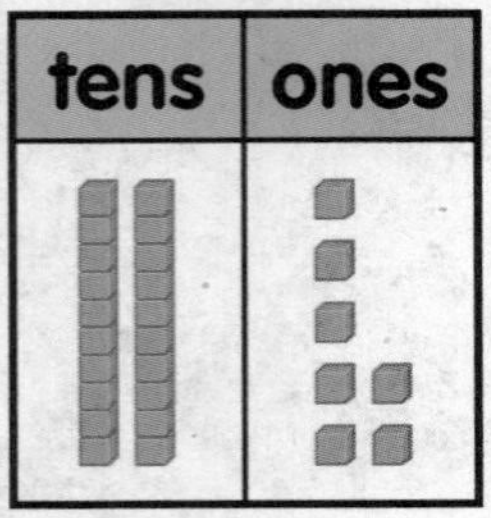

	tens	ones
	2	5
+		2

Step 2
Add the ones.

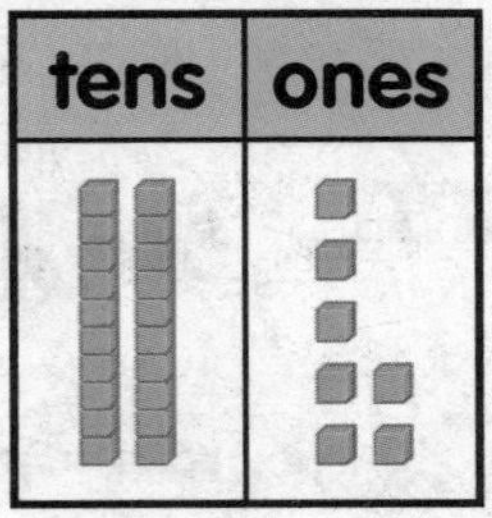

	tens	ones
	2	5
+		2
		7

Helpful Hint
Add the ones.
5 ones + 2 ones = 7 ones

Step 3
Add the tens.

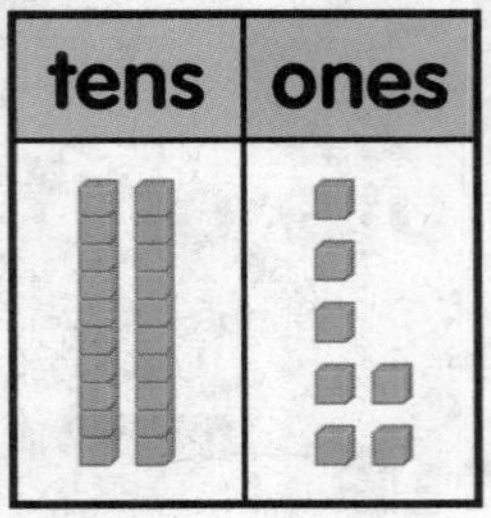

	tens	ones
	2	5
+		2
	2	7

The sum is 2 tens and 7 ones or 27.

Use Work Mat 7 and [ten rod] and [one cube]. Add.

1.

	tens	ones
	1	3
+		4

2.

	tens	ones
		6
+	4	0

3.

	tens	ones
		6
+	5	3

Talk Math Explain how you add tens and ones.

Name ....

Remember to start on the right.

On My Own

Use Work Mat 7 and  and . Add.

4.

	tens	ones
	2	2
+		5

5.

	tens	ones
	4	2
+		5

6.

	tens	ones
		4
+	4	4

7.

	tens	ones
	5	2
+		6

8.

	tens	ones
	7	1
+		4

9.

	tens	ones
		8
+	9	0

10.

	tens	ones
	1	4
+		4

11.

	tens	ones
	5	5
+		4

12.

	tens	ones
	8	2
+		4

13.

	tens	ones
	3	1
+		5

14.

	tens	ones
	7	2
+		6

15.

	tens	ones
	9	1
+		7

Problem Solving

Processes & Practices

16. 32 children are at the library. 4 more children come to the library. How many children are at the library in all?

_____ children

17. Oliver put together 53 puzzle pieces. He added 5 more pieces. How many puzzle pieces has Oliver put together in all?

_____ pieces

HOT Problem Which sum is more, 23 + 6 or 23 + 20? Explain.

Name ....

My Homework

Lesson 3
Add Tens and Ones

Homework Helper

eHelp Need help? connectED.mcgraw-hill.com

To find 43 + 6, add the ones. Then add the tens.

	tens	ones
	4	3
+		6
	4	9

Helpful Hint
Add the ones. 3 ones + 6 ones = 9 ones

The sum of 4 tens and 9 ones is 49.

Practice

Add.

1.

	tens	ones
	1	1
+		4

2.

	tens	ones
		6
+	6	0

3.

	tens	ones
	3	5
+		4

4.

	tens	ones
		4
+	2	4

5.

	tens	ones
	9	1
+		5

6.

	tens	ones
		7
+	5	2

Add.

7.

	tens	ones
	5	0
+		4

8.

	tens	ones
	4	5
+		4

9.

	tens	ones
	1	3
+		6

10.

	tens	ones
		1
+	8	7

11.

	tens	ones
	3	3
+		5

12.

	tens	ones
	9	3
+		4

13. There are 15 children rollerblading in the park. 3 more join them. How many children are rollerblading in all?

______ children

Test Practice

14.

	tens	ones
	7	2
+		3

75 ○ 76 ○ 77 ○ 78 ○

Math at Home Your child learned to add a two-digit number with a one-digit number. Ask your child to explain how to add 42 + 5.

Name ______________________

Problem Solving

STRATEGY: Guess, Check, and Revise

Lesson 4

ESSENTIAL QUESTION

How can I add and subtract two-digit numbers?

Taye sees two colors of bowling balls. He sees 27 bowling balls in all. Which two colors does he see?

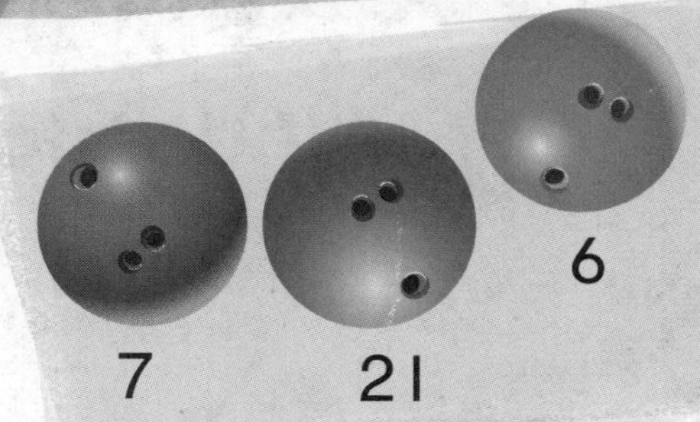

1 Understand

Underline what you know.
Circle what you need to find.

2 Plan

How will I solve the problem?

3 Solve

I will guess, check, and revise.

$21 + 7 = 28$ too many
$7 + 6 = 13$ not enough
$21 + 6 = 27$ correct

blue and pink bowling balls

4 Check

Is my answer reasonable? Explain.

Practice the Strategy

Allie has two different colors of marbles. There are 34 marbles in all. What two colors of marbles does she have?

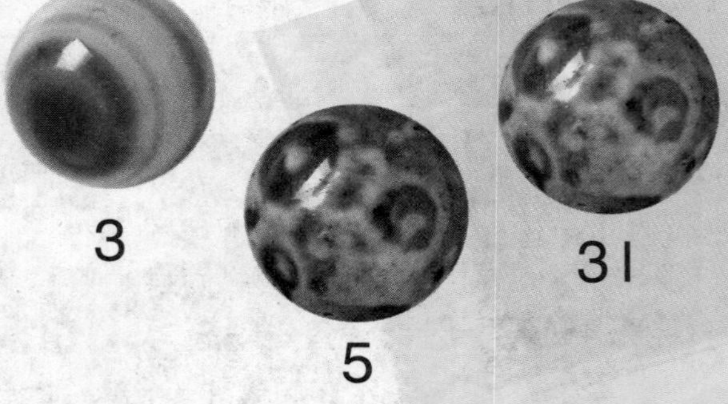

1 **Understand** Underline what you know.
Circle what you need to find.

2 **Plan** How will I solve the problem?

3 **Solve** I will...

__________ and __________ marbles

4 **Check** Is my answer reasonable? Explain.

Name ....

Apply the Strategy

1. Jerry has two colors of balloons. He has 19 balloons in all. What color of balloons does he have?

7 14 5

____

2. Tia finds two colors of leaves. She finds 46 leaves in all. What colors of leaves does she find?

42 4 7

____

3. Vito gets two toys. He uses 29 tickets to get them. What toys does he get?

5 6 7

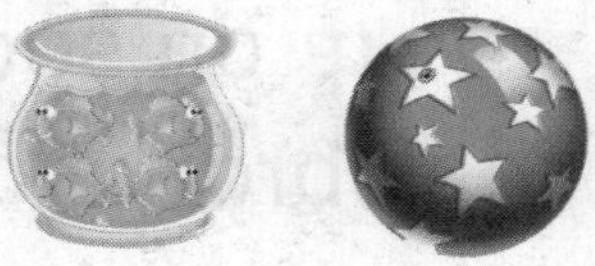

8 21

____

Review the Strategies

Choose a strategy

- Guess, check, and revise.
- Write a number sentence.
- Draw a diagram.

4. There are 2 children on the bus. 26 more children get on. How many children are on the bus altogether?

_____ children

5. There are 54 small dogs at a park. 4 large dogs are at the same park. How many dogs are at the park in all?

Will you take me to the park?

_____ dogs

6. A library has 4 books on soccer and 61 books on golf. How many books on soccer and golf does the library have?

_____ books

Name ....

My Homework

Lesson 4
Problem Solving: Guess, Check, and Revise

Homework Helper

eHelp Need help? connectED.mcgraw-hill.com

Ryan sees two colors of birds. He sees a total of 18 birds in all. Which two colors of birds does he see?

1 Understand Underline what you know. Circle what you need to find.

2 Plan How will I solve the problem?

3 Solve I will guess, check, and revise.

9 + 10 = 19 too many
9 + 8 = 17 not enough
10 + 8 = 18 correct

The colors of the birds are yellow and red.

4 Check Is my answer reasonable?

Problem Solving

Underline what you know.
Circle what you need to find.

1. Harper draws 12 pictures. Raven draws some pictures. Together they drew 15 pictures. How many pictures did Raven draw?

_____ pictures

2. There are 37 students in the band. The students play only two types of instruments. Each student plays one instrument. What two instruments are played? Circle them.

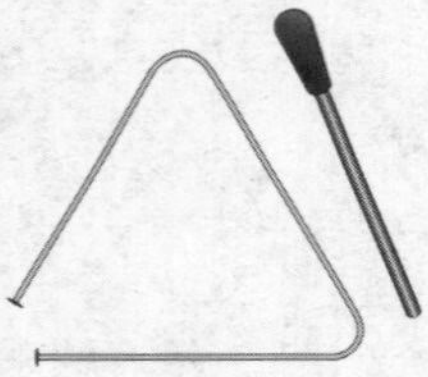
31

7

6

3. Viviana jumps rope 52 times. Josiah jumps rope 7 times. Adam jumps rope 6 times. What two children jump rope 59 times?

____

Math at Home Take advantage of problem-solving opportunities during daily routines such as riding in the car, bedtime, doing laundry, putting away groceries, planning schedules, and so on.

Name ....

Add Tens and Ones with Regrouping

Lesson 5

ESSENTIAL QUESTION
How can I add and subtract two-digit numbers?

Explore and Explain

Tools

tens	ones

19 + 3 = 22

Teacher Directions: Use [rod] and [unit] to model. Abby sees 19 people flying kites. Later, she sees 3 more people flying kites. How many people are flying kites in all? Trace how many. Draw the rods and units you used.

See and Show

Processes & Practices

Find 13 + 8.

Step 1
Count the ones.

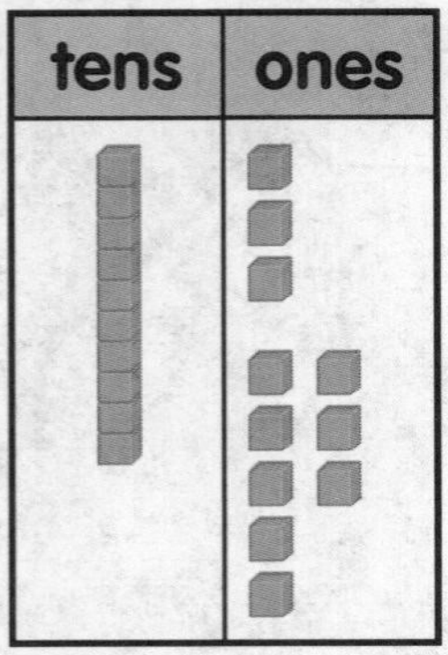

Step 2
Circle 10 ones.

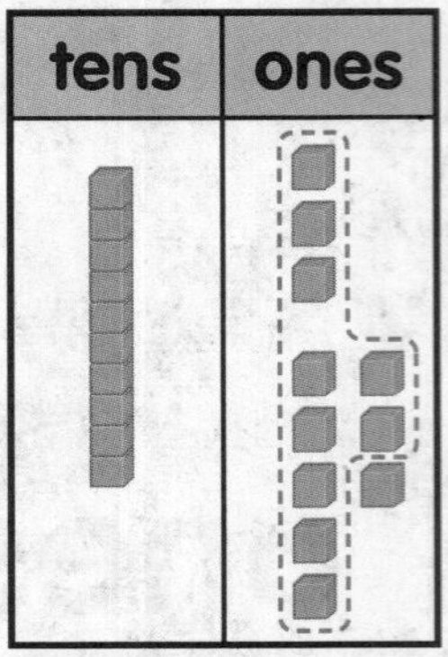

Step 3
Regroup 10 ones as 1 ten.

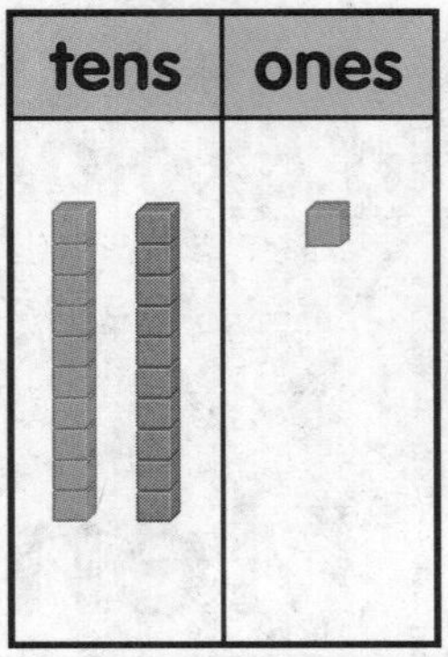

Helpful Hint
Move the ones to the tens column.
10 ones = 1 ten

__2__ tens and __1__ one is 21.

The sum of 18 + 3 is 21.

Circle the ones to show regrouping. Write your answer.

1. 15 + 5 = ______

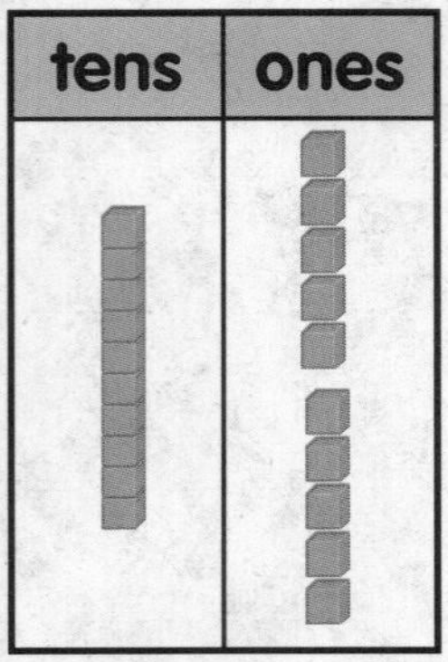

2. 19 + 4 = ______

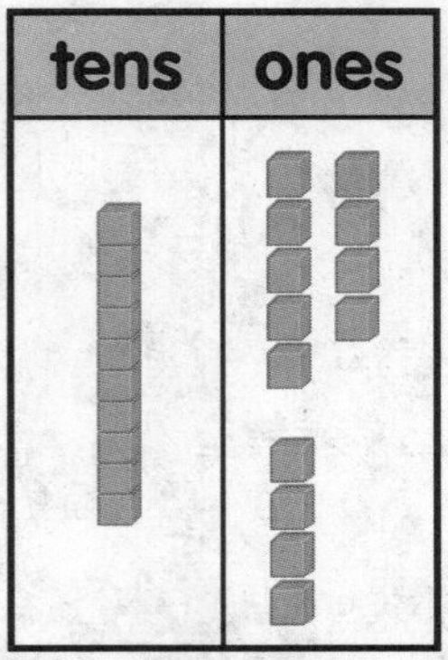

Talk Math Do you always regroup when adding? Explain.

Name ....

On My Own

Circle the ones to show regrouping. Write your answer.

3. 13 + 8 = ______

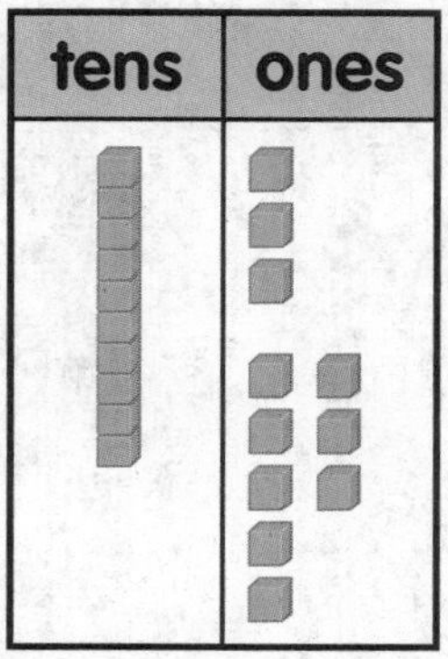

4. 18 + 7 = ______

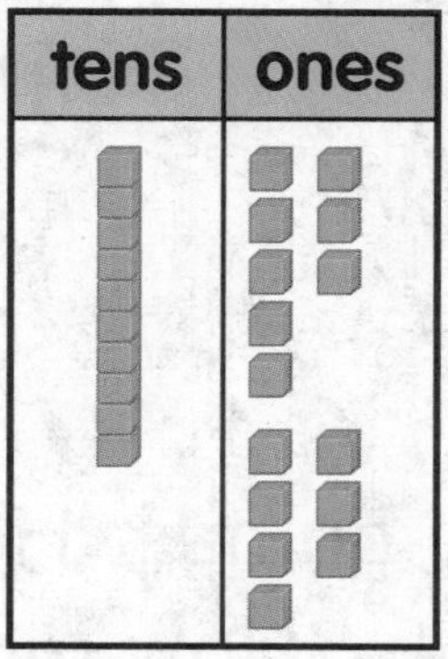

5. 19 + 7 = ______

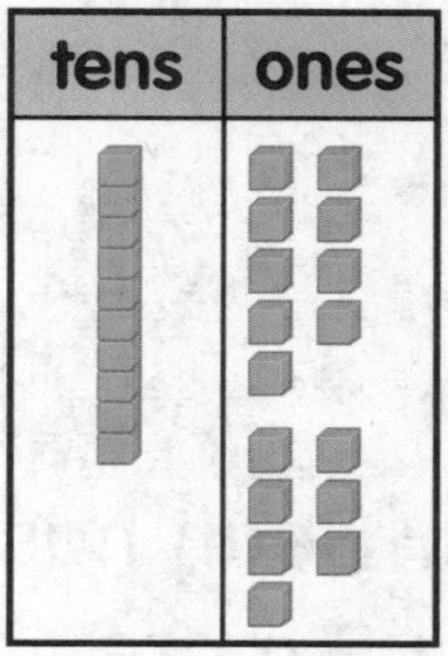

6. 16 + 5 = ______

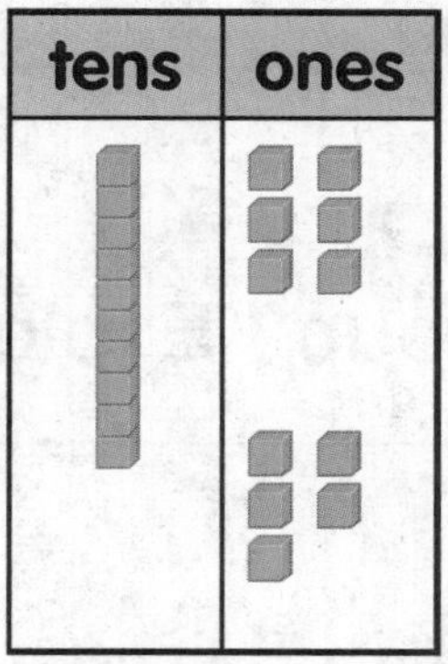

7. 18 + 9 = ______

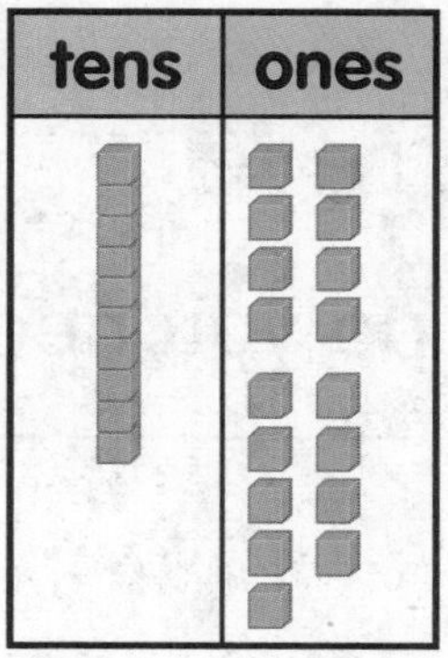

8. 12 + 8 = ______

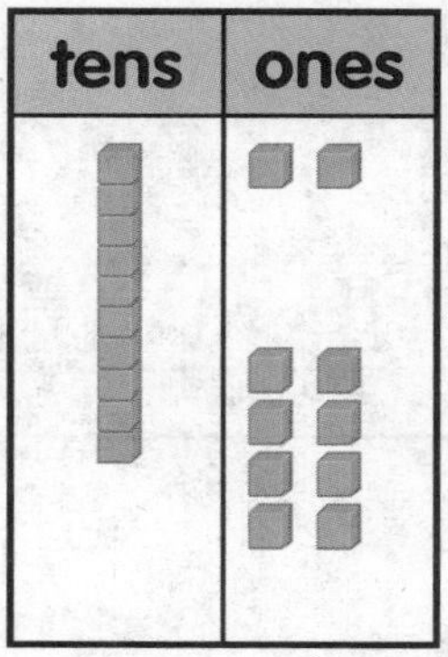

9. 17 + 7 = ______

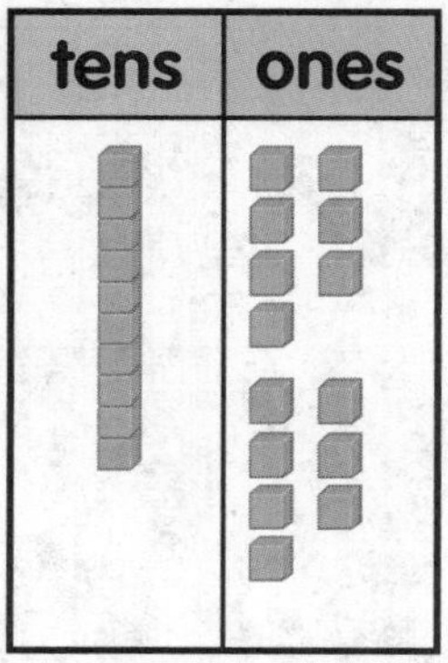

10. 19 + 9 = ______

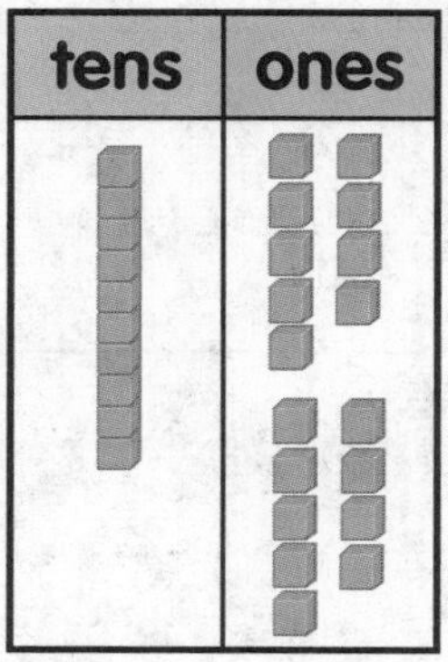

11. 16 + 7 = ______

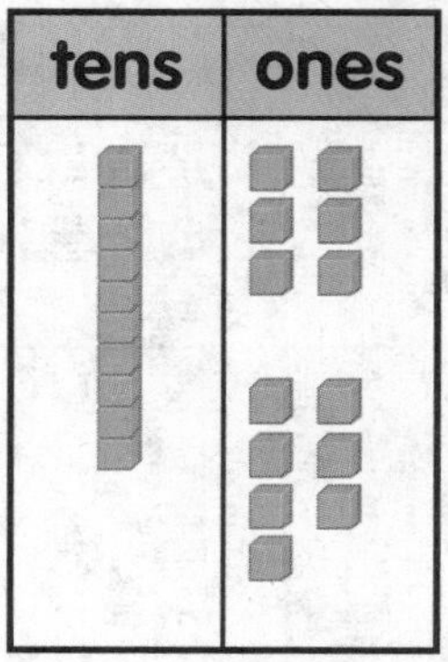

Real World!

Problem Solving

Processes & Practices

12. Mrs. Brown's class went to the zoo. 16 students saw the lions. 8 other students saw the monkeys. How many students went to the zoo in all?

_____ students

13. Sean has 14 video games. He is given 9 more video games. How many video games does Sean have in all?

_____ video games

Write Math When do you need to regroup? Explain.

____

____

____

Name ____________________

My Homework

Lesson 5

Add Tens and Ones with Regrouping

Homework Helper

eHelp Need help? connectED.mcgraw-hill.com

Find 16 + 7.

Step 1

Count the ones.

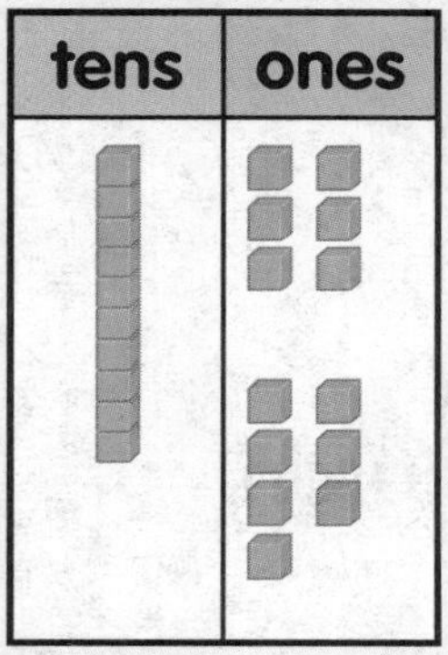

Step 2

Circle 10 ones.

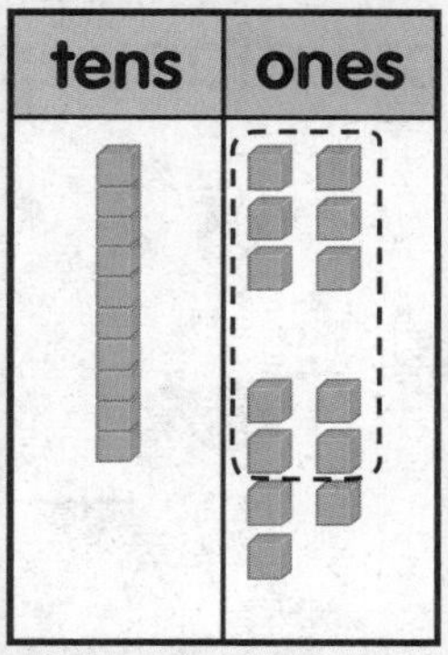

Step 3

Regroup 10 ones as 1 ten.

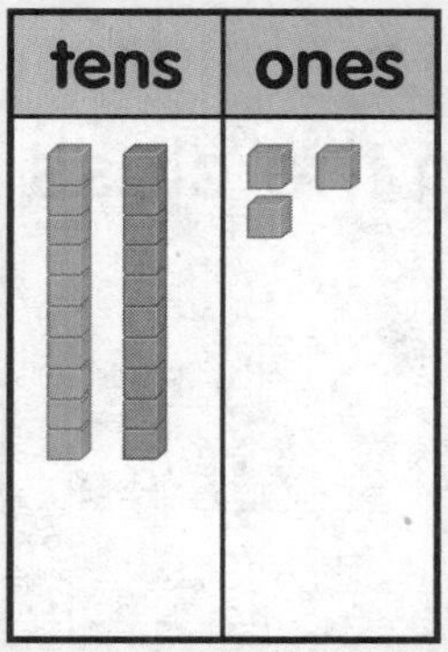

So, 16 + 7 = 23.

Practice

Circle the ones to show regrouping. Write your answer.

1. 19 + 7 = ______

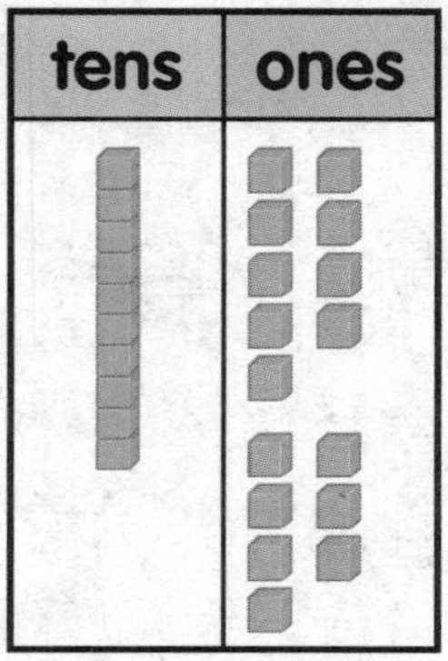

2. 15 + 8 = ______

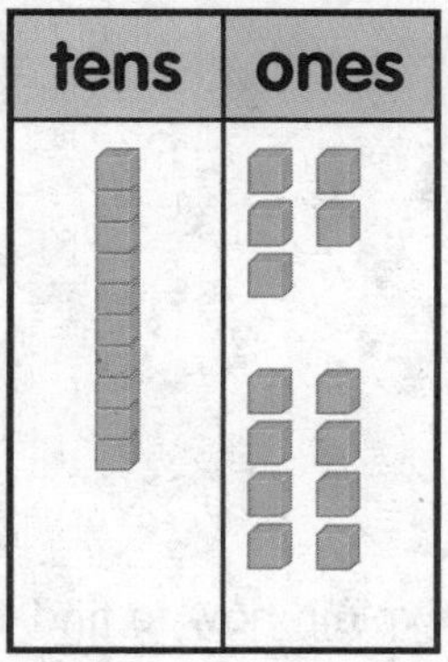

3. 14 + 9 = ______

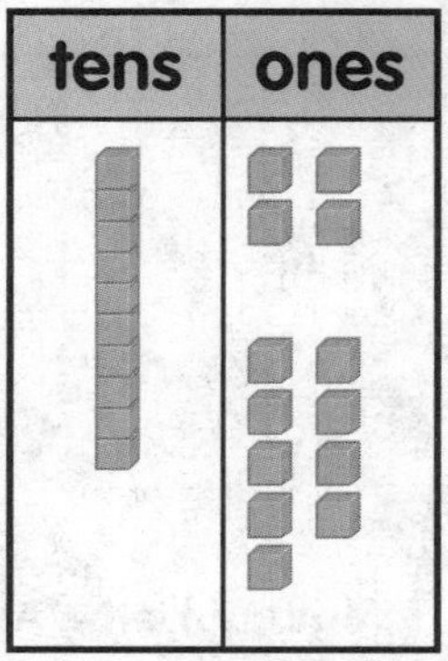

Circle the ones to show regrouping. Write your answer.

4. $19 + 5 =$ ______

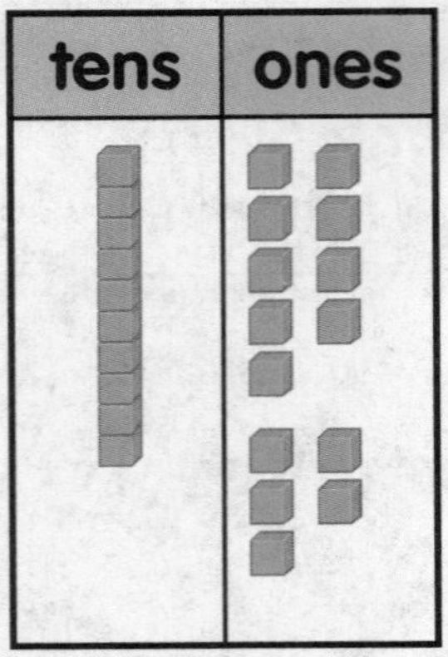

5. $13 + 7 =$ ______

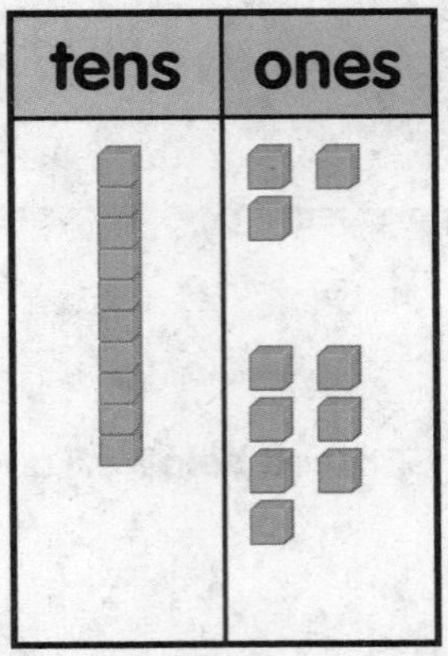

6. $16 + 6 =$ ______

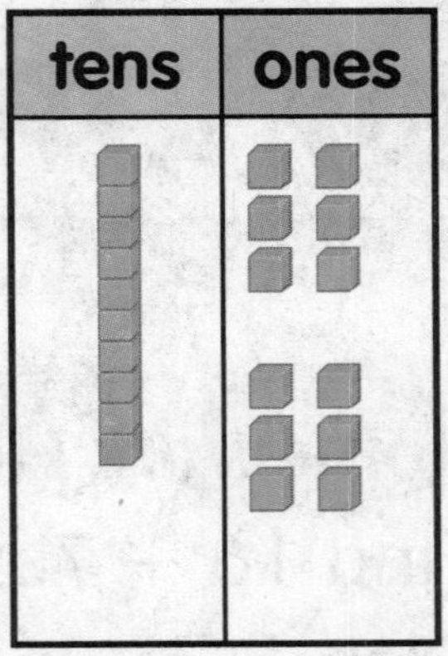

7. There are 16 children in a ballet class. 5 more children join. How many children are in the class now?

______ children

Test Practice

8. Find $15 + 6$.

○ 21 ○ 20 ○ 19 ○ 15

Math at Home Ask your child to explain how to find $12 + 9$.

Name ____________________

Check My Progress

Vocabulary Check

Complete each sentence.

adding **subtracting** **sum**

1. By ______________ two numbers together, you can find the sum.
2. The answer to an addition problem is called the ______________.

Concept Check

Add.

3. 3 tens + 2 tens = _____ tens 30 + 20 = _____

4. 4 tens + 4 tens = _____ tens 40 + 40 = _____

5. 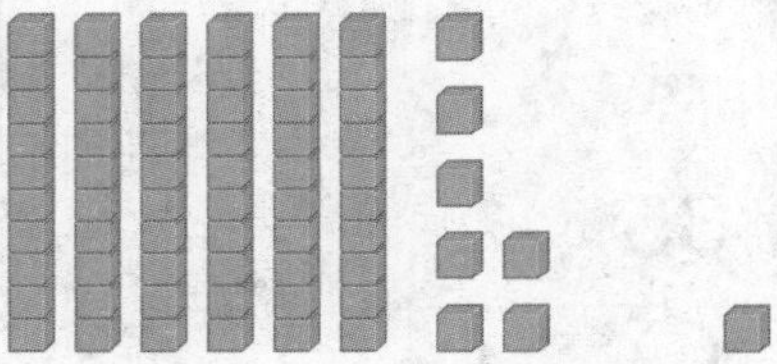

67 + 1 = _____

6. 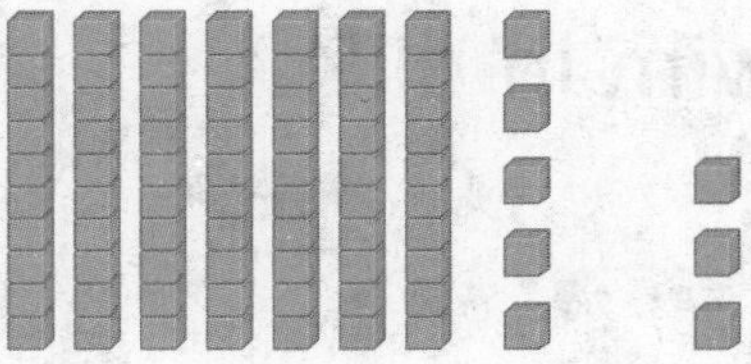

75 + 3 = _____

Add.

7.

	tens	ones
	3	5
+		4

8.

	tens	ones
	5	2
+		7

9.

	tens	ones
	8	4
+		5

Circle the ones to show regrouping.
Write your answer.

10. 16 + 8 = ____

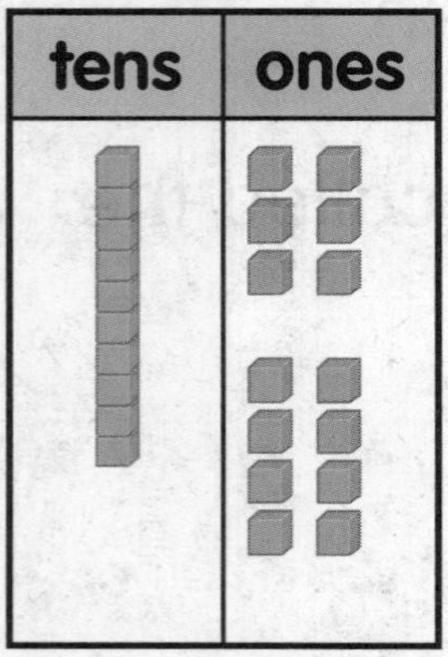

11. 14 + 8 = ____

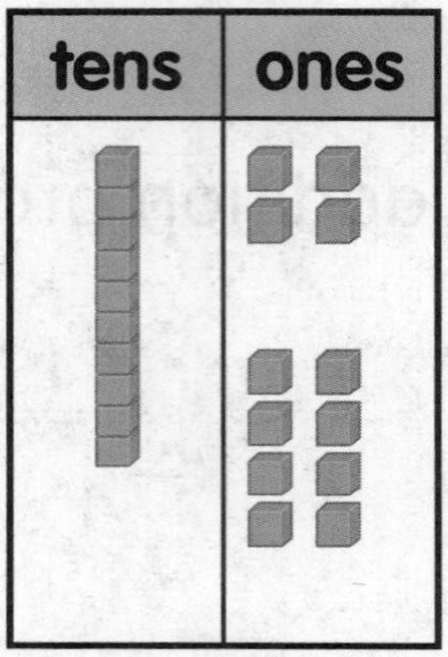

12. 18 + 9 = ____

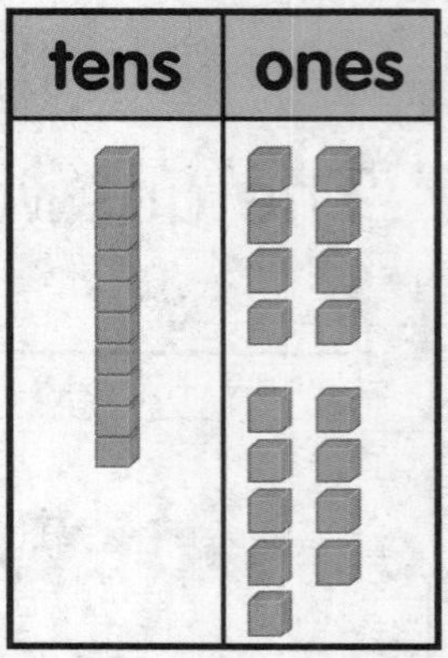

Test Practice

13. Dalton swam 25 laps in the swimming pool. Madison swam 9 laps. How many laps did they swim in all?

16 ○ 25 ○ 34 ○ 35 ○

Name

Subtract Tens

Lesson 6

ESSENTIAL QUESTION
How can I add and subtract two-digit numbers?

Explore and Explain

Watch Tools

_____ lightning bugs

Teacher Directions: Use [ten-rod] to model. Alex put 40 lightning bugs in a jar. He let 20 of the lightning bugs go. How many bugs are left in the jar? Write the number. Draw the rods to show your work. Mark an X on the rods to show the bugs that were let go.

See and Show

Processes & Practices

Find 30 − 20.

3 tens − 2 tens = 1 ten

30 − 20 = 10

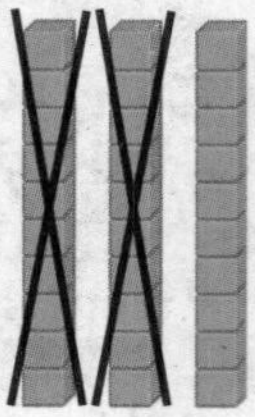

Helpful Hint
To subtract 30 − 20, subtract the tens.

Find 50 − 10.

5 tens − 1 ten = 4 tens

50 − 10 = 40

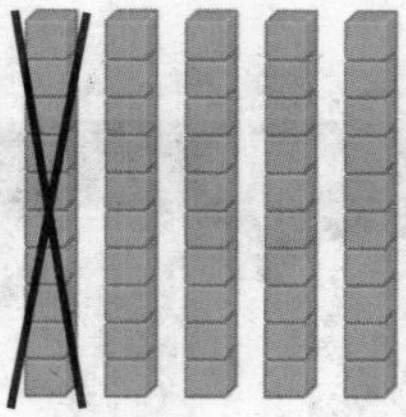

5 tens − 1 ten = 4 tens or 40.

Subtract. Use  to help.

1. 4 tens − 2 tens = ______ tens   40 − 20 = ______

2. 6 tens − 1 ten = ______ tens   60 − 10 = ______

3. 7 tens − 3 tens = ______ tens   70 − 30 = ______

4. 9 tens − 3 tens = ______ tens   90 − 30 = ______

Talk Math How does knowing 6 − 2 help you find 60 − 20?

Name ______________________

On My Own

Subtract. Use [ten rod] to help.

5. 2 tens − 2 tens = ______ tens   20 − 20 = ______

6. 8 tens − 3 tens = ______ tens   80 − 30 = ______

7. 9 tens − 5 tens = ______ tens   90 − 50 = ______

8. 7 tens − 1 ten = ______ tens   70 − 10 = ______

9. 4 tens − 3 tens = ______ ten   40 − 30 = ______

10. 9 tens − 2 tens = ______ tens   90 − 20 = ______

11. 5 tens − 3 tens = ______ tens   50 − 30 = ______

12. 6 tens − 5 tens = ______ ten   60 − 50 = ______

13. 8 tens − 6 tens = ______ tens   80 − 60 = ______

Processes & Practices

Problem Solving

14. Justin caught 30 fish. Paul caught 20 fish. How many more fish did Justin catch than Paul?

_____ fish

15. There are 40 children in Lily's dance class. 20 of them are girls. How many of them are boys?

_____ boys

HOT Problem 80 people are in line to ride a roller coaster. 20 people get on the roller coaster. The answer is 60 people. What is the question?

Name ____________________

My Homework

Lesson 6
Subtract Tens

Homework Helper

eHelp

Need help? connectED.mcgraw-hill.com

Find 80 − 40.

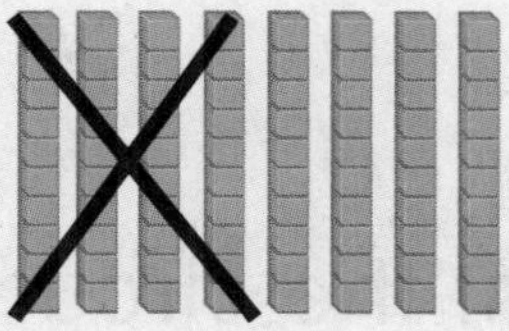

8 tens − 4 tens = 4 tens
80 − 40 = 40

Helpful Hint
To subtract 80 − 40, subtract the tens.

Practice

Subtract.

1. 4 tens − 3 tens = ______ ten 40 − 30 = ______

2. 7 tens − 2 tens = ______ tens 70 − 20 = ______

3. 6 tens − 6 tens = ______ tens 60 − 60 = ______

4. 9 tens − 2 tens = ______ tens 90 − 20 = ______

5. 8 tens − 6 tens = ______ tens 80 − 60 = ______

Subtract.

6.

8 tens	80
− 7 tens	− 70
ten	

7.

5 tens	50
− 1 ten	− 10
tens	

8.

6 tens	60
− 3 tens	− 30
tens	

9.

9 tens	90
− 3 tens	− 30
tens	

10. There are 50 children at the circus.
20 of those children leave to go home.
How many children are still at the circus?

______ children

Test Practice

11. There are 70 people at a movie theatre.
20 of them are eating popcorn. How many people are not eating popcorn?

30 people ○ 50 people ○ 60 people ○ 70 people ○

Math at Home Have your child tell you how many tens are left in 70 − 40.

Name ______________________________

Count Back by 10s

Lesson 7

ESSENTIAL QUESTION
How can I add and subtract two-digit numbers?

Explore and Explain

Tools

70 − 30 = ______

Teacher Directions: Use [connecting cubes icon] to model. Samuel starts at 70. He hops and lands on 70 − 30. Draw a circle around the number where he lands. Write the number. Explain to a classmate how you found the answer.

See and Show

Processes & Practices

Use a number line to subtract numbers by tens.

Helpful Hint
Start at 50. Count back by tens to find the difference. 40, 30, 20, 10.
0 10 20 30 40 50 60 70 80 90 100

50 − 40 = 10

Use the number line to subtract. Show your work. Write the difference.

1. 80 − 30 = ______

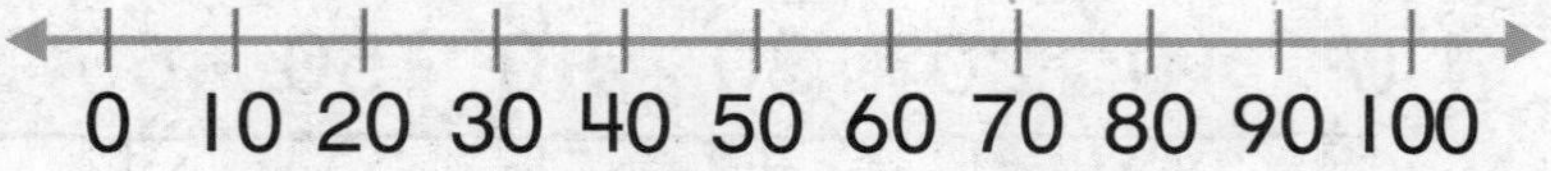

2. 50 − 20 = ______

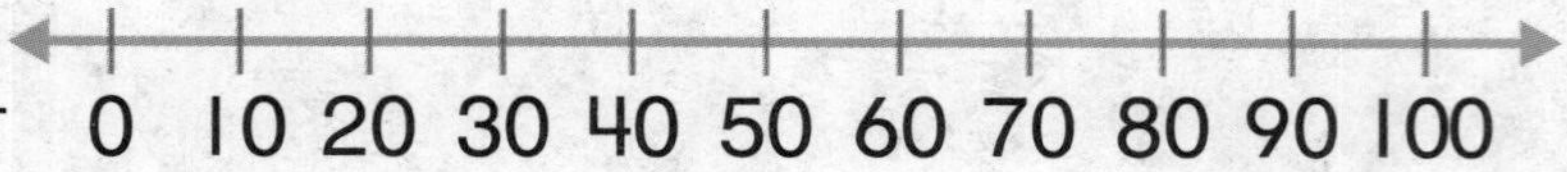

Talk Math Explain how you can use a number line to help you subtract by tens.

Name ....

On My Own

Use the number line to help you subtract.
Write the difference.

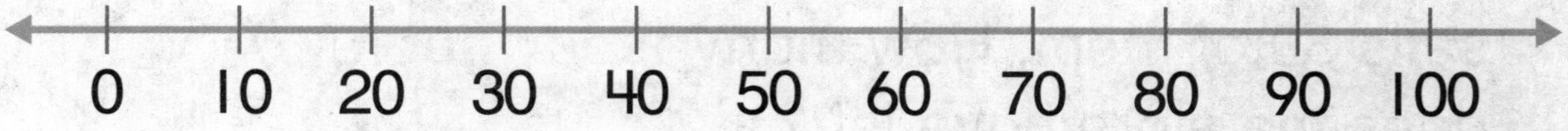

3. 40 − 20 = ____

4. 60 − 20 = ____

5. 80 − 70 = ____

6. 20 − 10 = ____

7. 90 − 40 = ____

8. 50 − 20 = ____

9. 70 − 50 = ____

10. 30 − 10 = ____

11. 80 − 40 = ____

12. 60 − 30 = ____

13. 40 − 30 = ____

14. 50 − 10 = ____

15. 90 − 20 = ____

Real World! Problem Solving

Processes & Practices

Use the number line to help you subtract.

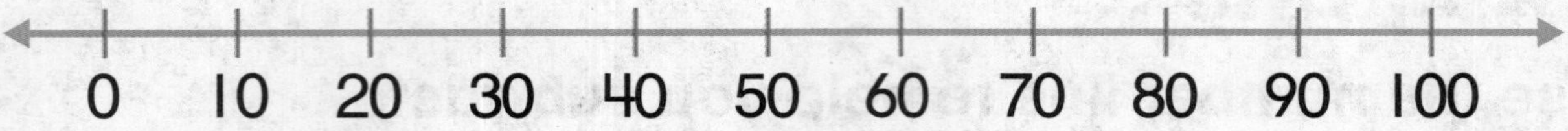

16. A toy store has 40 teddy bears. The store sells 30 of them. How many teddy bears does the store have left?

______ teddy bears

17. Sadie was given a set of 70 blocks. She lost 40 of the blocks. How many blocks does Sadie have left?

______ blocks

Write Math How can you use a number line to find 30 – 30? What is the difference?

Name ....

My Homework

Lesson 7

Count Back by 10s

Homework Helper

eHelp

Need help? connectED.mcgraw-hill.com

To find 70 − 30, start at 70.

Count back by tens on a number line.

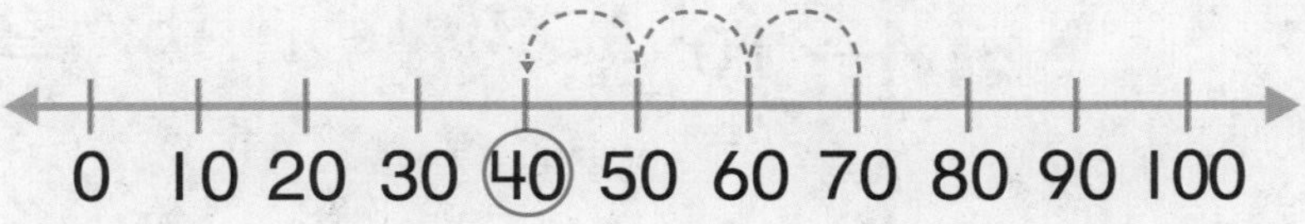

70 − 30 = 40

Practice

Use the number line to subtract. Show your work. Write the difference.

1. 60 − 20 = ______

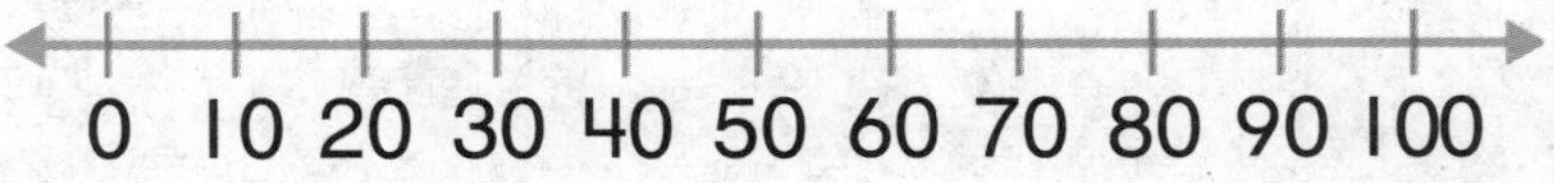

2. 90 − 10 = ______

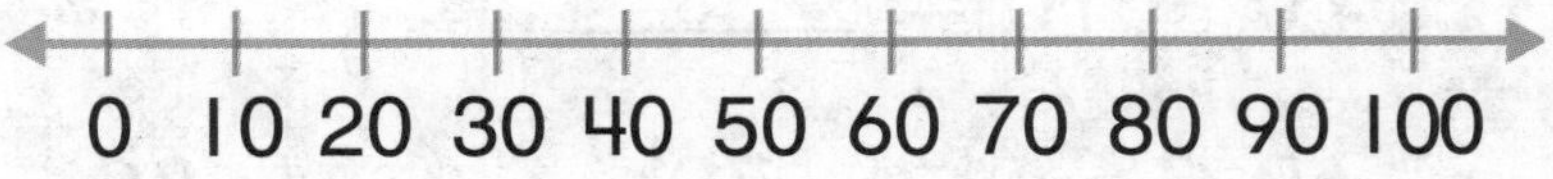

3. 80 − 30 = ______

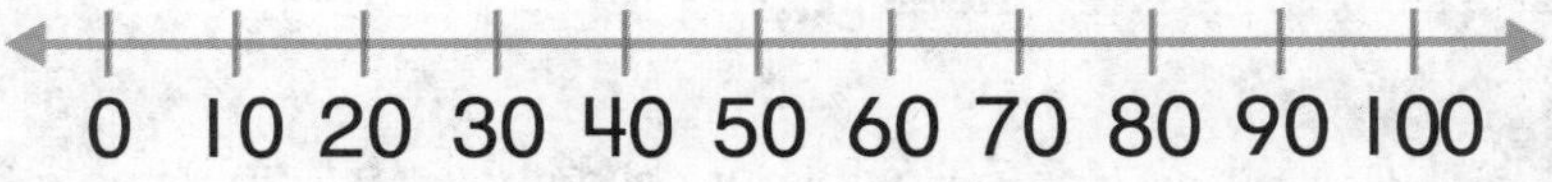

Use the number line to help you subtract.
Write the difference.

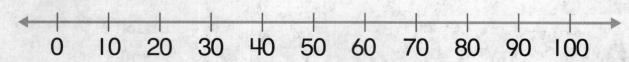

0 10 20 30 40 50 60 70 80 90 100

4. 50 − 40 = ______

5. 80 − 10 = ______

6. 30 − 10 = ______

7. 70 − 20 = ______

8. 90 − 30 = ______

9. 40 − 10 = ______

10. 60 − 40 = ______

11. There are 30 children in line to go down a slide. 10 children go down the slide. How many children are still in line?

______ children

Test Practice

12. 70 − 10 = ______

○ 60 ○ 70 ○ 80 ○ 90

Math at Home Have your child show 60 − 20 using a number line. Have him or her explain how they use the number line as they subtract.

Name ______________________________

Relate Addition and Subtraction of Tens

Lesson 8

ESSENTIAL QUESTION
How can I add and subtract two-digit numbers?

Explore and Explain

Tools

50 + 30 = _____

_____ − _____ = _____

Teacher Directions: Use [base-ten rods] to model. There are 50 people ice skating. 30 more people join them. How many people are ice skating in all? Write the number. Write a related subtraction problem. Explain how the number sentences are related.

See and Show

Processes & Practices

Find 70 – 20.

To find 70 – 20, think 20 + 50 = 70.

So, 70 – 20 = 50.

Helpful Hint
20 + 50 = 70,
so 70 – 20 = 50.

Use related facts to add and subtract.

1. 10 + _____ = 30

   30 – 10 = _____

2. 30 + _____ = 60

   60 – 30 = _____

3. 40 + _____ = 50

   50 – 40 = _____

4. 20 + _____ = 40

   40 – 20 = _____

Subtract. Write a related addition fact.

5. 80 – 50 = _____

   _____ + _____ = _____

6. 90 – 70 = _____

   _____ + _____ = _____

Talk Math Tell the related addition fact you would use to find 60 – 40. Explain.

Name ________________

Remember to use related facts.

On My Own

Use related facts to add and subtract.

7. 50 + ____ = 70

70 − 50 = ____

8. 20 + ____ = 80

80 − 20 = ____

9. 40 + ____ = 90

90 − 40 = ____

10. 50 + ____ = 60

60 − 50 = ____

11. 20 + ____ = 50

50 − 20 = ____

12. 30 + ____ = 70

70 − 30 = ____

Subtract. Write a related addition fact.

13. 40 − 10 = ____

____ + ____ = ____

14. 90 − 10 = ____

____ + ____ = ____

15. 30 − 20 = ____

____ + ____ = ____

16. 90 − 50 = ____

____ + ____ = ____

Problem Solving

Processes & Practices

Use [tens rod] to subtract.
Write a related addition fact.

17. A toy store has 60 video games. 40 of them get sold. How many video games does the store have left?

____ − ____ = ____   ____ + ____ = ____

18. Marcus has 90 basketball cards. He gives 20 cards to his brother. How many cards does Marcus have left?

____ − ____ = ____   ____ + ____ = ____

Write Math Write a word problem using the number sentence 20 + 30 = 50 or 50 − 20 = 30.

Name ____________________

My Homework

Lesson 8

Relate Addition and Subtraction of Tens

Homework Helper

eHelp **Need help?** connectED.mcgraw-hill.com

Related facts can help you find missing numbers.

To find 70 − 50, think 50 + 20 = 70.

So, 70 − 50 = 20

Practice

Use related facts to add and subtract.

1. 10 + ____ = 20

   20 − 10 = ____

2. 40 + ____ = 60

   60 − 40 = ____

3. 50 + ____ = 90

   90 − 50 = ____

4. 20 + ____ = 30

   30 − 20 = ____

5. 30 + ____ = 80

   80 − 30 = ____

6. 10 + ____ = 50

   50 − 10 = ____

Subtract. Write a related addition fact.

7. 70 − 60 = ____

____ + ____ = ____

8. 40 − 20 = ____

____ + ____ = ____

Write the subtraction number sentence.
Then write a related addition fact.

9. Ming's team scores 50 points in their first basketball game and 30 points in their second game. How many more points did they score in their first game?

____ − ____ = ____ ____ + ____ = ____

Test Practice

Mark the related subtraction fact.

10. 40 + ____ = 60

- 40 + 10 = 60 ○
- 60 + 30 = 90 ○
- 60 − 40 = 20 ○
- 60 − 40 = 10 ○

Math at Home Have your child solve the following subtraction fact: 80 − 60 = ___. Ask them to say the related addition fact that can help them solve the problem.

Name ______________________________

My Review

Chapter 6

Two-Digit Addition and Subtraction

Vocabulary Check

Vocab

Complete each sentence.

difference	**ones**	
subtract	**sum**	**tens**

1. The answer to an addition problem is called the ______________.

2. In the number 63, 6 is in the ______________ place.

3. You ________________ to find the difference.

4. The answer to a subtraction problem is called the ________________.

5. In the number 72, 2 is in the ______________ place.

Concept Check

Add.

6. 2 tens + 1 ten = _____ tens   20 + 10 = _____

7. 6 tens + 3 tens = _____ tens   60 + 30 = _____

Count on to add. Write the sum.

8.

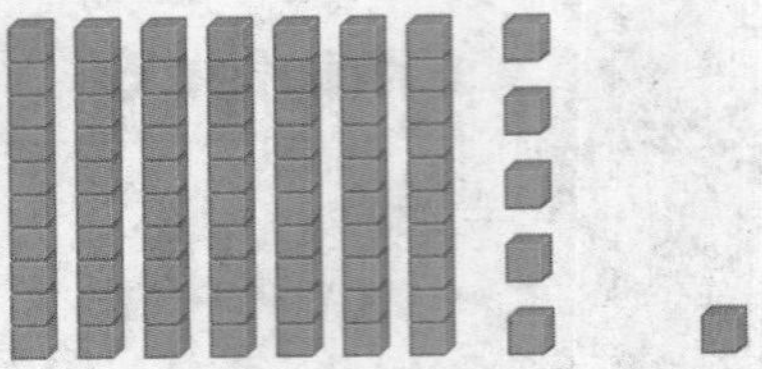

75 + 1 = ______

9.

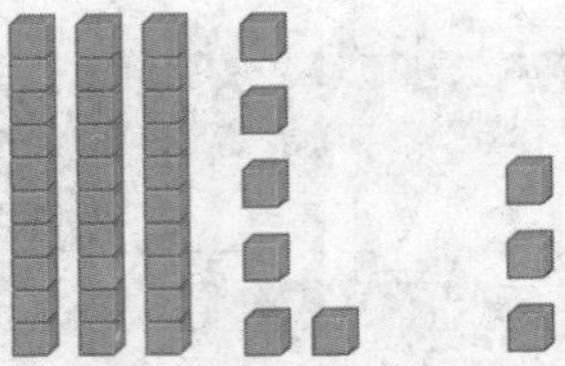

36 + 3 = ______

Add.

10.

	tens	ones
	6	2
+		3

11.

	tens	ones
	3	7
+		2

12.

	tens	ones
	5	4
+		4

Subtract.

13. 6 tens − 2 tens = ______ tens 60 − 20 = ______

14. 5 tens − 2 tens = ______ tens 50 − 20 = ______

Use the number line to help you subtract.

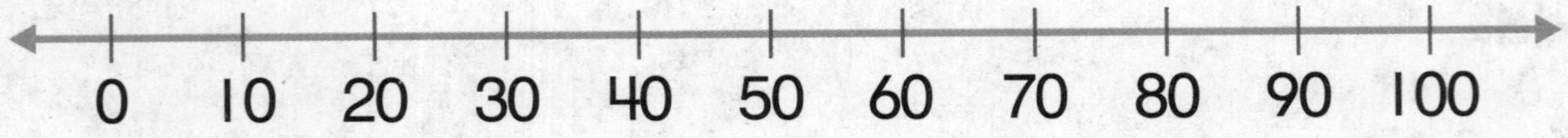

15. 80 − 30 ______

16. 90 − 20 ______

Name

Problem Solving

Use the number line to help you subtract.

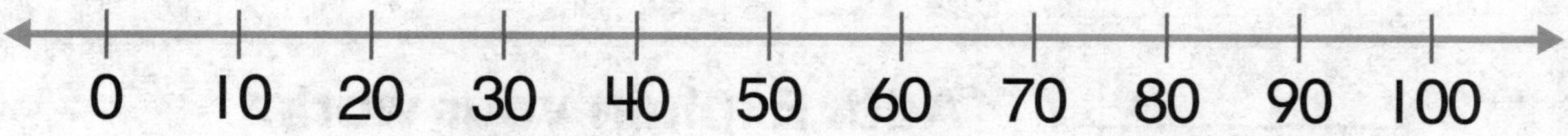

17. Cameron has 60 marbles. Eric borrows 30 of them. How many marbles does Cameron have left?

______ marbles

Subtract. Write the related addition fact.

18. Sam drew 40 pictures. He gave 20 of them to his dad. How many pictures does Sam have left?

_____ − _____ = _____   _____ + _____ = _____

Test Practice

19. There are 25 people at a park. 3 more people come to the park. How many people are at the park in all?

22 people	27 people	28 people	55 people
○	○	○	○

Reflect

Chapter 6

Answering the Essential Question

ESSENTIAL QUESTION

How can I add and subtract two-digit numbers?

Add. Explain your work.

	tens	ones
	6	2
+		3

Subtract. Explain your work.

8 tens − 3 tens = ______ tens

80 − 30 = ______

Now I Know!!!

Glossary/Glosario

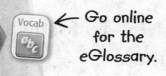

Go online for the eGlossary.

Aa

English | **Spanish/Español**

add (adding, addition) To join together sets to find the total or sum.

2 + 5 = 7

sumar (adición) Unir conjuntos para hallar el total o la suma.

2 + 5 = 7

addend Any numbers or quantities being added together.

2 + 3
↑ ↑

2 is an addend and 3 is an addend.

sumando Números o cantidades que se suman.

2 + 3
↑ ↑

2 es un sumando y 3 es un sumando.

addition number sentence An expression using numbers and the + and = signs.

4 + 5 = 9

enunciado numérico de suma Expresión en la cual se usan números con los signos + e =.

4 + 5 = 9

Aa

after To follow in place or time.

5 6 7 8

6 is just *after* 5

después Que sigue en lugar o en tiempo.

5 6 7 8

6 está justo *después* de 5.

analog clock A clock that has an hour hand and a minute hand.

reloj analógico Reloj que tiene manecilla horaria y minutero.

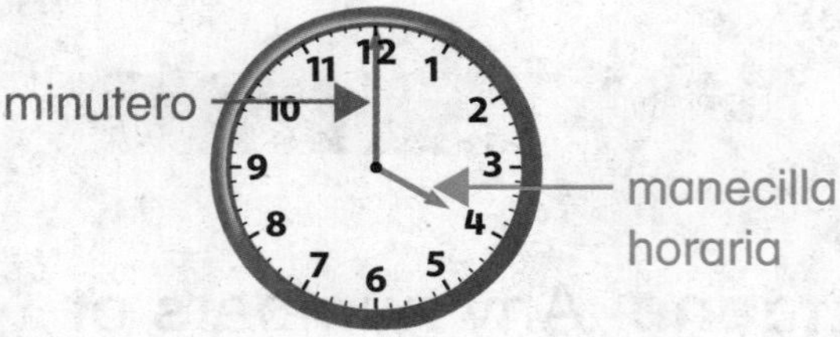

Bb

bar graph A graph that uses bars to show data.

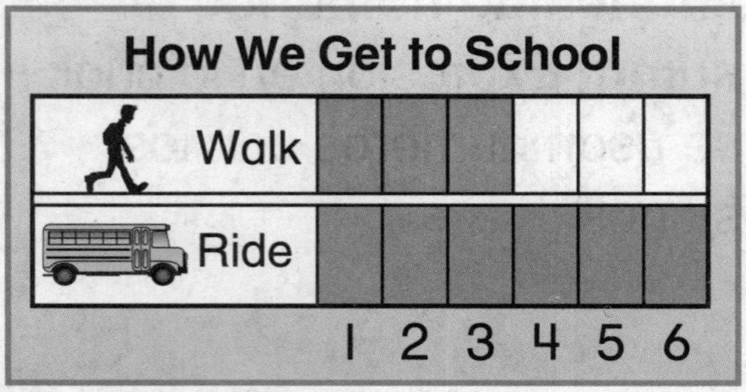

gráfica de barras Gráfica que usa barras para ilustrar datos.

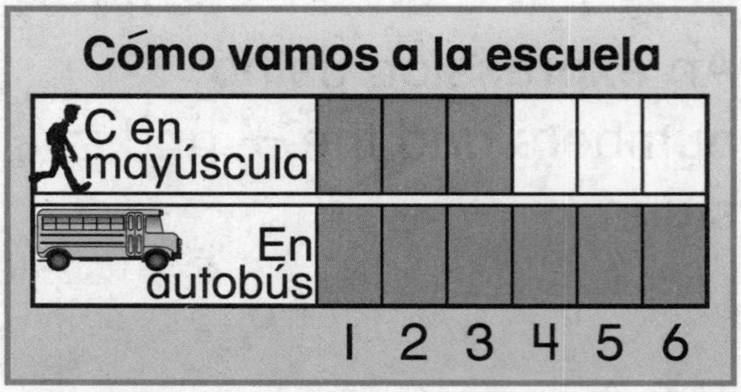

before

5 6 7 8

6 is just *before* 7.

antes

5 6 7 8

6 está justo *antes* del 7.

between

The kitten is *between* the two dogs.

entre

El gatito está *entre* dos perros.

Cc

capacity The amount of dry or liquid material a container can hold.

capacidad Cantidad de material seco o líquido que cabe en un recipiente.

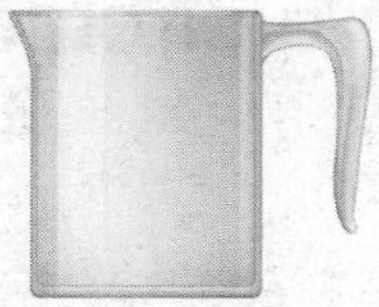

Cc

cent ¢

1¢ 1 cent

centavo ¢

1¢ 1 centavo

circle A closed round shape.

círculo Figura redonda y cerrada.

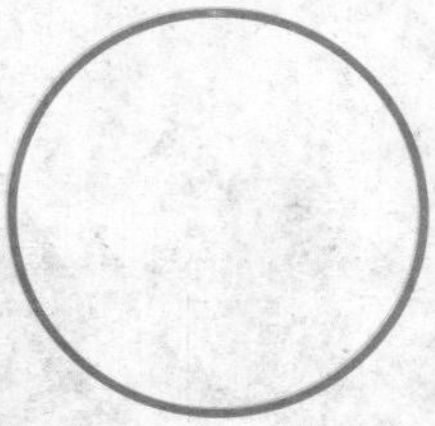

compare Look at objects, shapes, or numbers and see how they are alike or different.

comparar Observar objetos, formas o números para saber en qué se parecen y en qué se diferencian.

cone A three-dimensional shape that narrows to a point from a circular face.

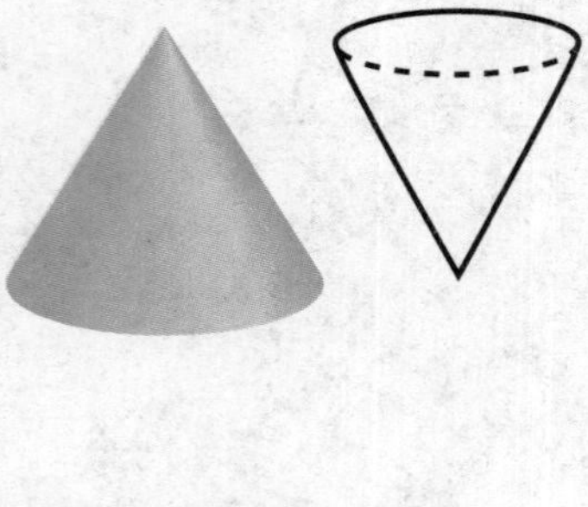

cono Una figura tridimensional que se estrecha hasta un punto desde una cara circular.

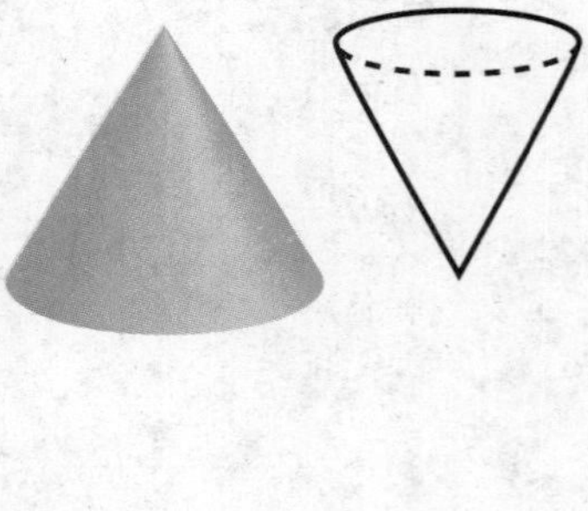

count back On a number line, start at the number 5 and count back 3.

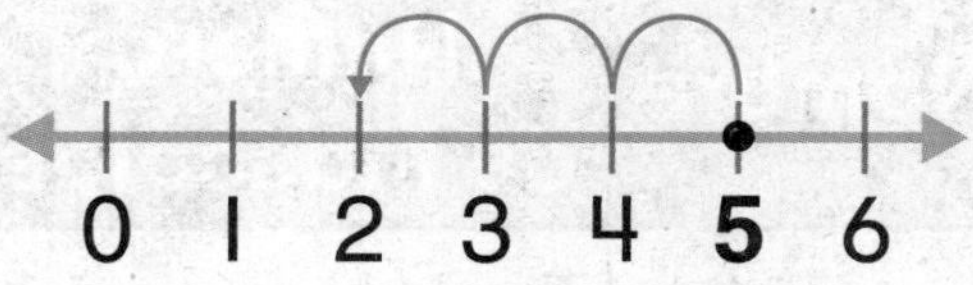

5 − 3 = 2 Count back 3.

contar hacia atrás En una recta numérica, comienza en el número 5 y cuenta 3 hacia atrás.

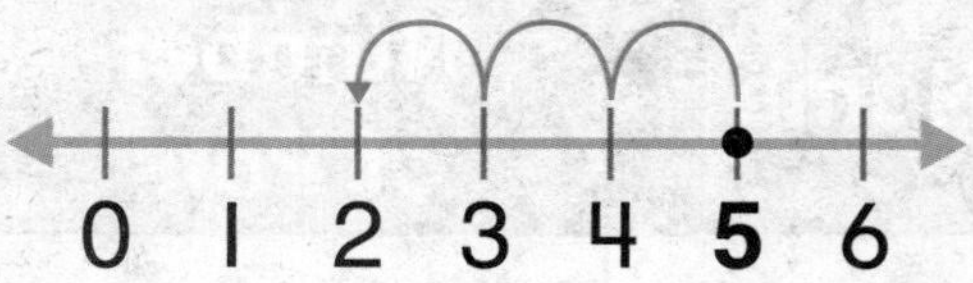

5 − 3 = 2 Cuenta 3 hacia atrás.

count on (or count up) On a number line, start at the number 4 and count up 2.

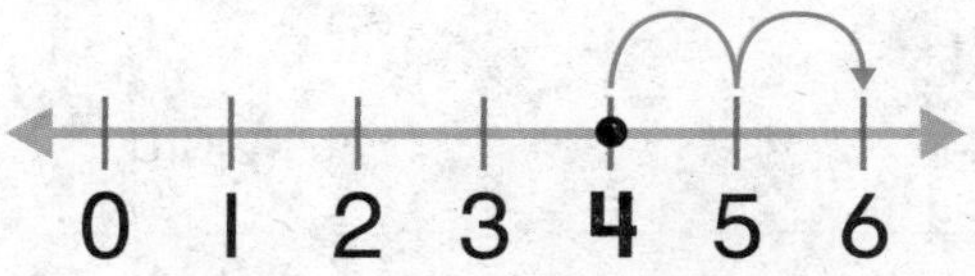

4 + 2 = 6 Count on 2.

seguir contando (o contar hacia delante) En una recta numérica, comienza en el 4 y cuenta 2.

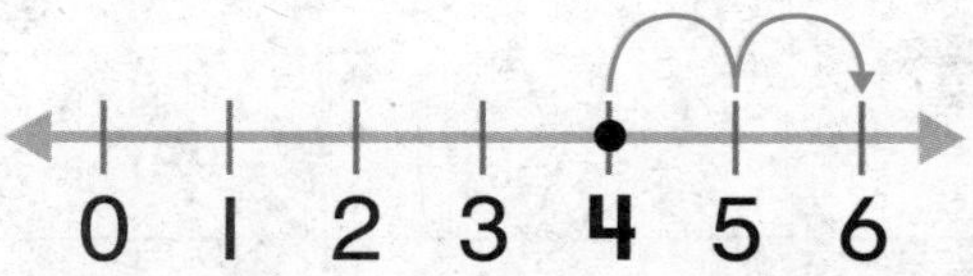

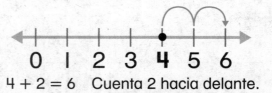

4 + 2 = 6 Cuenta 2 hacia delante.

cube A three-dimensional shape with 6 square faces.

cubo Una figura tridimensional con 6 caras cuadradas.

cylinder A three-dimensional shape that is shaped like a can.

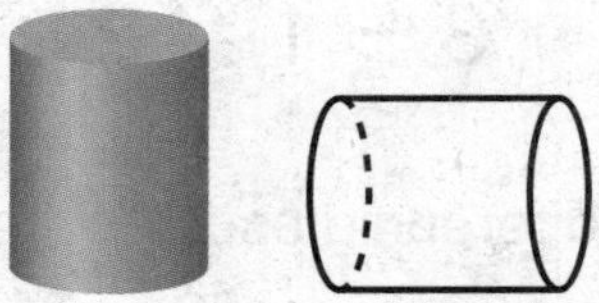

cilindro Una figura tridimensional que tiene la forma de una lata.

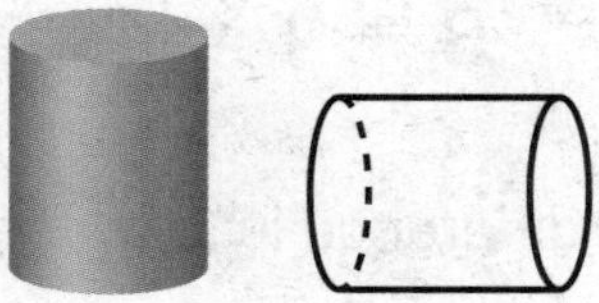

Dd

data Numbers or symbols collected to show information.

Name	Number of Pets
Mary	3
James	1
Alonzo	4

datos Números o símbolos que se recopilan para mostrar información.

Nombre	Número de mascotas
Maria	3
James	1
Alonzo	4

day

día

difference The answer to a subtraction problem.

$$3 - 1 = 2$$

The difference is 2.

diferencia Resultado de un problema de resta.

$$3 - 1 = 2$$

La diferencia es 2.

digital clock A clock that uses only numbers to show time.

dime dime = 10¢ or 10 cents

head tail

doubles (doubles plus 1, near doubles) Two addends that are the same number.

2 + 2 = 4

2 + 3 = 5 2 + 1 = 3

reloj digital Reloj que usa solo números para mostrar la hora.

moneda de 10¢ moneda de diaz centavos = 10¢ o 10 centavos

cara cruz

dobles (y dobles más 1, casi dobles) Dos sumandos que son el mismo número.

2 + 2 = 4

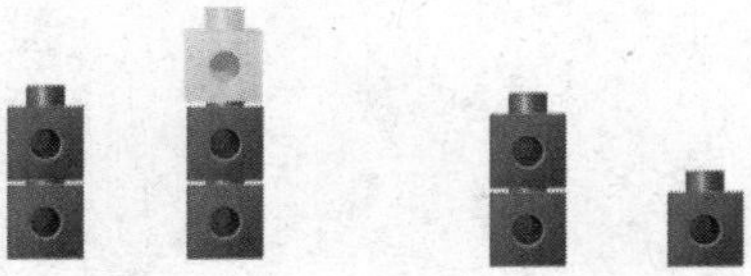

2 + 3 = 5 2 + 1 = 3

Ee

equal parts Each part is the same size.

A muffin cut in equal parts.

partes iguales Cada parte es del mismo tamaño.

Un panecillo cortado en partes iguales.

equal to =

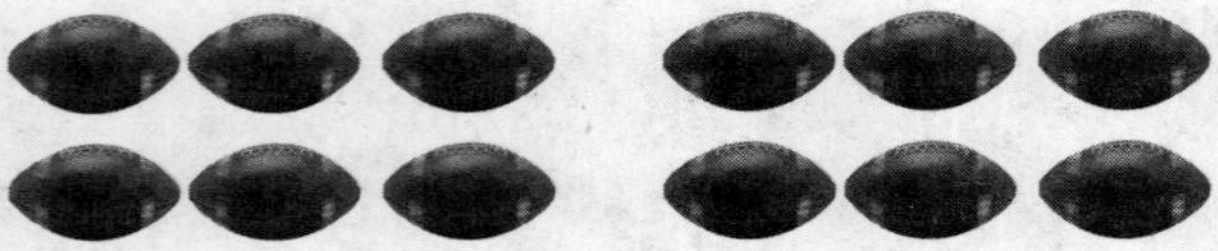

6 = 6

6 is equal to 6.

igual a =

6 = 6

6 es igual a 6.

equals (=) Having the same value as or is the same as.

2 + 4 = 6

equals sign ↑

igual (=) Que tienen el mismo valor o son lo mismo.

2 + 4 = 6

signo igual ↑

Ff

face The flat part of a three-dimensional shape.

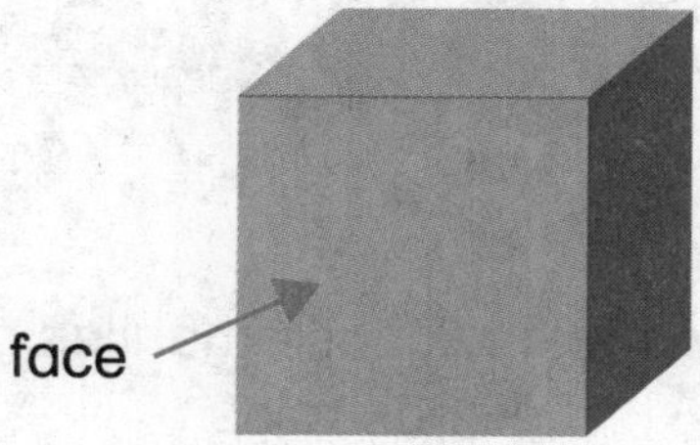

fact family Addition and subtraction sentences that use the same numbers. Sometimes called *related facts*.

$6 + 7 = 13$ $13 - 7 = 6$

$7 + 6 = 13$ $13 - 6 = 7$

false Something that is not a fact. The opposite of true.

cara La parte plana de una figura tridimensional.

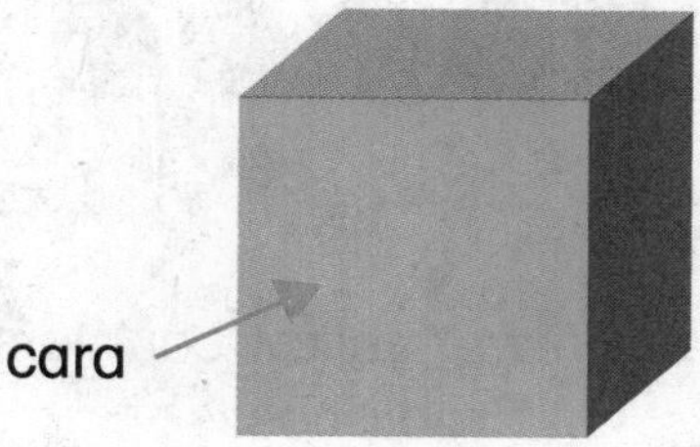

familia de operaciones Enunciados de suma y resta que tienen los mismos números. Algunas veces se llaman *operaciones relacionadas*.

$6 + 7 = 13$ $13 - 7 = 6$

$7 + 6 = 13$ $13 - 6 = 7$

falso Algo que no es cierto. Lo opuesto de verdadero.

Ff

fewer/fewest The number or group with less.

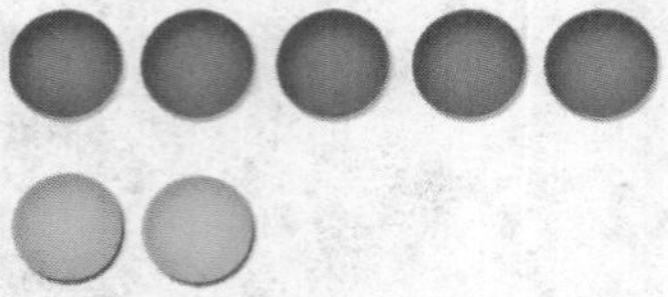

There are fewer yellow counters than red ones.

menos/el menor El número o grupo con menos.

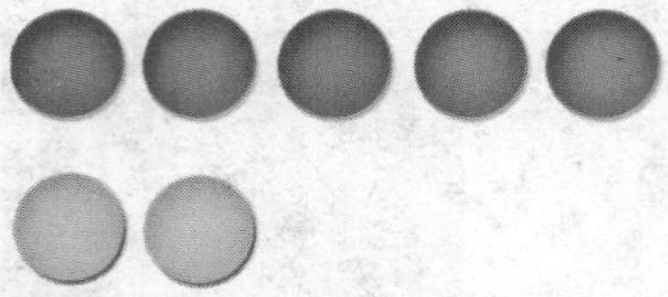

Hay menos fichas amarillas que fichas rojas.

fourths Four equal parts of a whole. Each part is a fourth, or a quarter of the whole.

cuartos Cuatro partes iguales de un todo. Cada parte es un cuarto, o la cuarta parte del todo.

Gg

graph A way to present data collected.

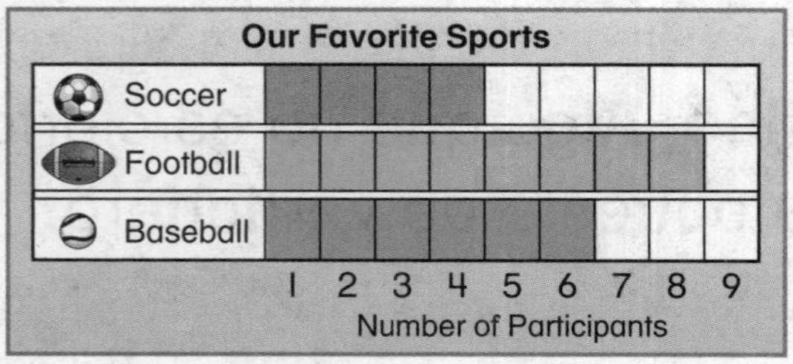

bar graph

gráfica Forma de presentar datos recopilados.

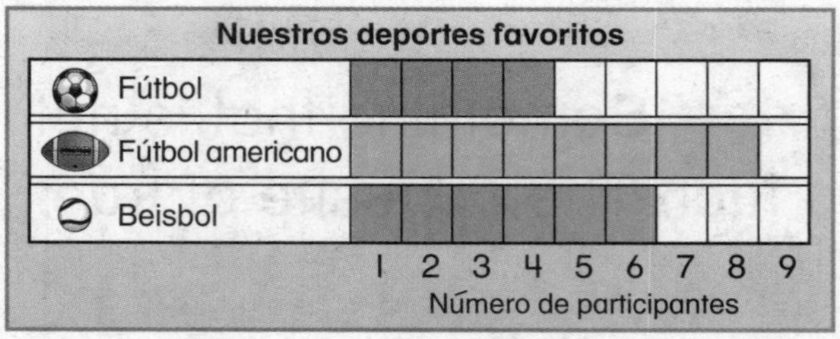

gráfica de barras

greater than (>)/greatest The number or group with more.

4	23	56

56 is the greatest.

mayor que (>)/el mayor El número o grupo con más cantidad.

4	23	56

56 es el mayor.

Hh

half hour (or half past)
One half of an hour is 30 minutes. Sometimes called *half past* or *half past the hour.*

media hora (o y media)
Media hora son 30 minutos. A veces se dice *hora y media.*

halves Two equal parts of a whole. Each part is a half of the whole.

mitades Dos partes iguales de un todo. Cada parte es la mitad de un todo.

heavy (heavier, heaviest)
Weighs more.

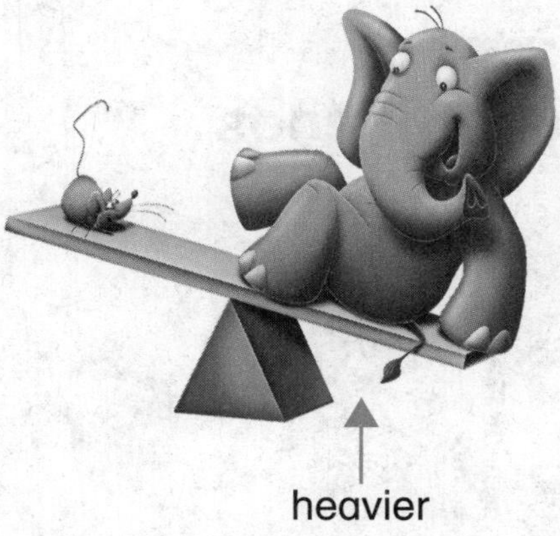

An elephant is heavier than a mouse.

pesado (más pesado, el más pesado) Pesa más.

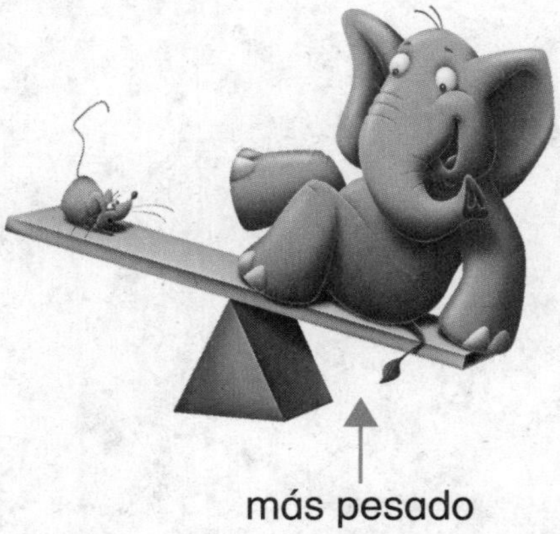

Un elefante es más pesado (pesa más) que un ratón.

Hh

height

altura

hexagon A two-dimensional shape that has six sides.

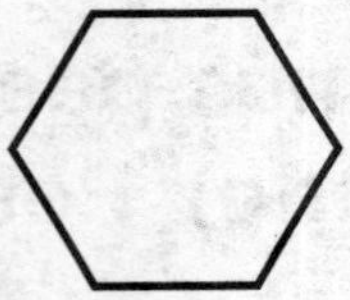

hexágono Figura bidimensional que tiene seis lados.

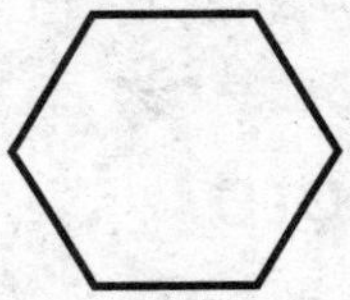

holds less/least

The glass holds less than the pitcher.

contener menos

El vaso contiene menos que la jarra.

holds more/most

The pitcher holds more than the glass.

contener más

La jarra contiene más que el vaso.

hour

hour A unit of time.

1 hour = 60 minutes

hora

hora Unidad de tiempo.

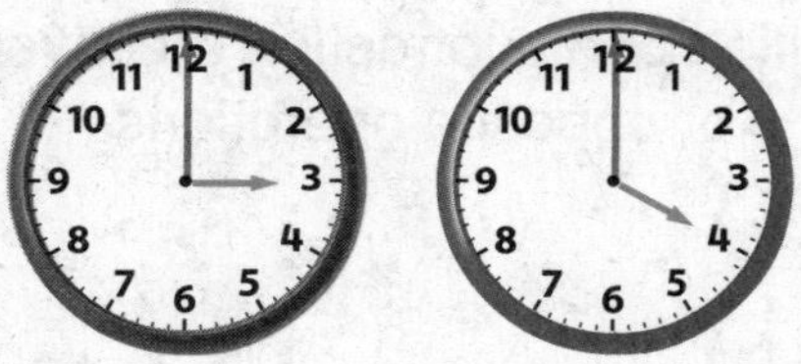

1 hora = 60 minutos

hour hand

hour hand The hand on a clock that tells the hour. It is the shorter hand.

manecilla horaria

manecilla horaria Manecilla del reloj que indica la hora. Es la manecilla más corta.

Hh

hundreds The numbers in the range of 100-999. It is the place value of a number.

centenas Los números en el rango del 100 al 999. Es el valor posicional de un número.

Ii

inverse Operations that undo each other.

Addition and subtraction are inverse or opposite operations.

operaciones inversas Operaciones que se anulan entre sí.

La suma y la resta son operaciones inversas u opuestas.

Ll

length

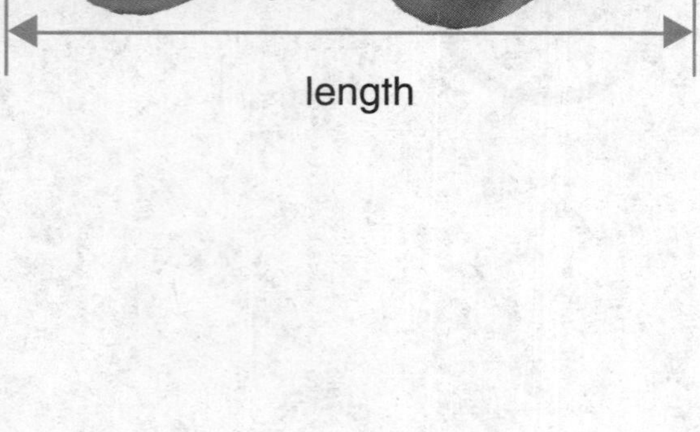

length

longitud

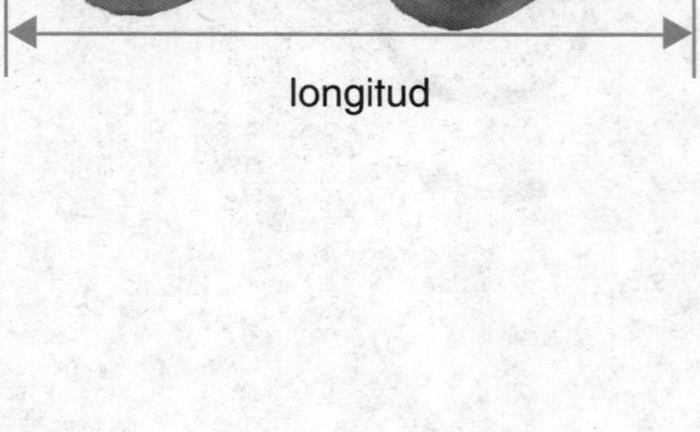

longitud

less than (<)/least The number or group with fewer.

4	23	56

4 is the least.

menor que (<)/el menor El número o grupo con menos cantidad.

4	23	56

4 es el menor.

light (lighter, lightest) Weighs less.

The mouse is lighter than the elephant.

liviano (más liviano, el más liviano) Pesa menos.

El ratón es más liviano (pesa menos) que el elefante.

long (longer, longest) A way to compare the lengths of two objects.

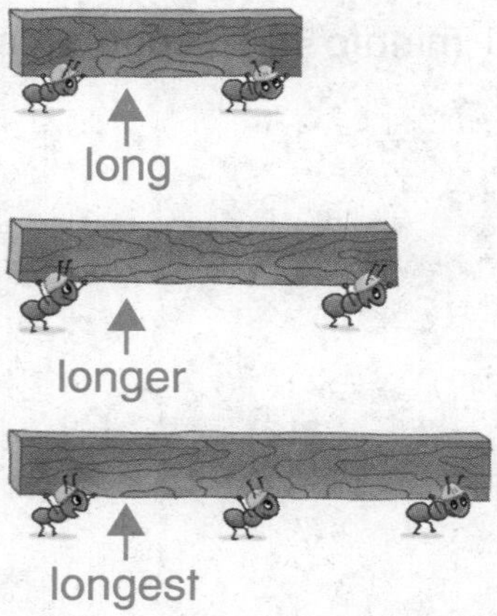

largo (más largo, el más largo) Forma de comparar la longitud de dos objetos.

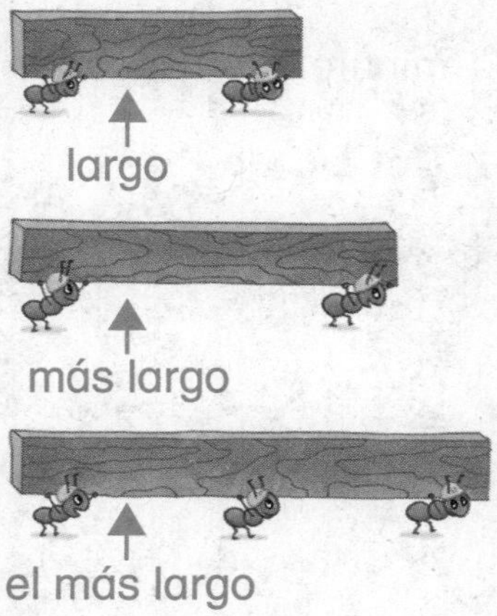

Mm

mass The amount of matter in an object. The mass of an object never changes.

masa Cantidad de materia en un objeto. La masa de un cuerpo nunca cambia.

measure To find the length, height, weight or capacity using standard or nonstandard units.

medir Hallar la longitud, altura, peso o capacidad mediante unidades estándar o no estándar.

minus (−) The sign used to show subtraction.

$$5 - 2 = 3$$

minus sign

menos (−) Signo que indica resta.

$$5 - 2 = 3$$

signo menos

minute (min) A unit to measure time.

1 minute = 60 seconds

minuto (min) Unidad que se usa para medir el tiempo.

1 minuto = 60 segundos

minute hand The longer hand on a clock that tells the minutes.

minutero La manecilla más larga del reloj. Indica los minutos.

missing addend

9 + ____ = 16

The missing addend is 7.

sumando desconocido

9 + ____ = 16

El sumando desconocido es 7.

month

mes

mes
abril
domingo lunes martes miércoles jueves viernes sábado
1 2 3 4 5
6 7 8 9 10 11 12
13 14 15 16 17 18 19
20 21 22 23 24 25 26
27 28 29 30

Mm

more

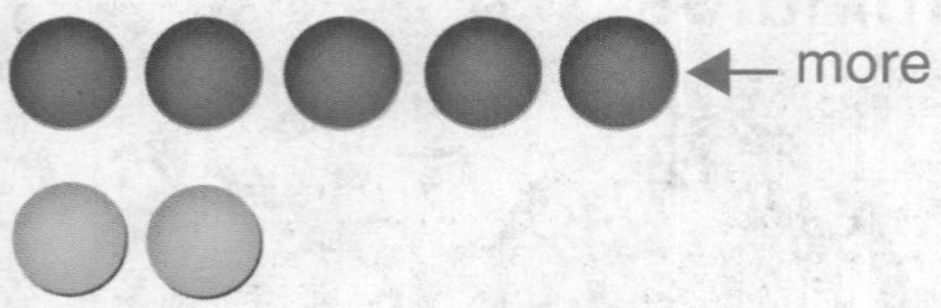

más

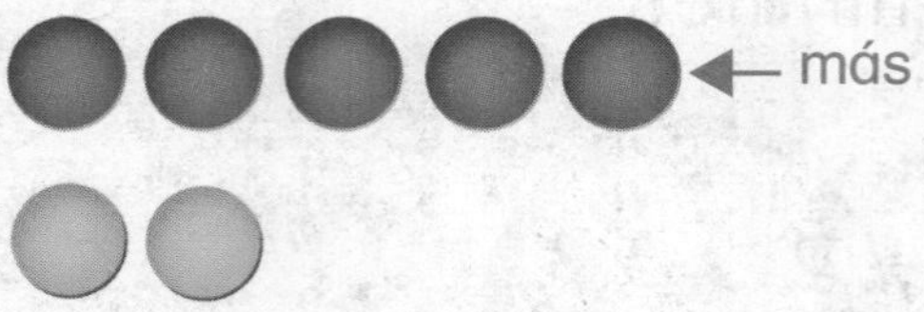

Nn

nickel nickel = 5¢ or 5 cents

head tail

moneda de 5¢ moneda de cinco centavos = 5¢ o 5 centavos

cara cruz

number Tells how many.
1, 2, 3, 4, 5, 6, 7, 8, 9, 10 ...

There are 3 chicks.

número Dice cuántos hay.
1, 2, 3, 4, 5, 6, 7, 8, 9, 10 ...

Hay tres pollitos.

number line A line with number labels.

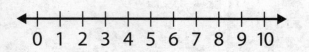

recta numérica Recta con marcas de números.

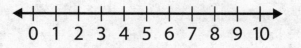

Oo

o'clock At the beginning of the hour.

It is 3 o'clock.

en punto Al comienzo de la hora.

Son las 3 en punto.

ones The numbers in the range of 0–9. It is the place value of a number.

unidades Los números en el rango de 0 a 9. Es el valor posicional de un número.

order

1, 3, 6, 7, 9

These numbers are in order from least to greatest.

orden

1, 3, 6, 7, 9

Estos números están en orden del menor al mayor.

ordinal number

first second third

númeral ordinal

primero segundo tercero

Pp

part One of the parts joined when adding.

Part	Part
2	2
Whole	

parte Una de las partes que se juntan al sumar.

Parte	Parte
2	2
El total	

pattern An order that a set of objects or numbers follows over and over.

A, A, B, A, A, B, A, A, B

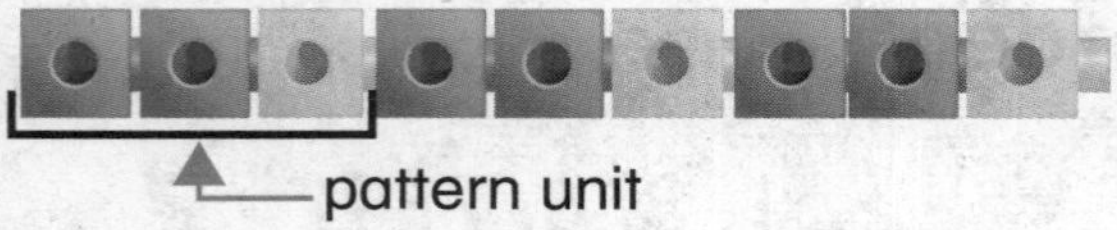

patrón Orden que sigue continuamente un conjunto de objetos o números.

A, A, B, A, A, B, A, A, B

penny penny = 1¢ or 1 cent

moneda de 1¢ moneda de un centavo = 1¢ o 1 centavo

picture graph A graph that has different pictures to show information collected.

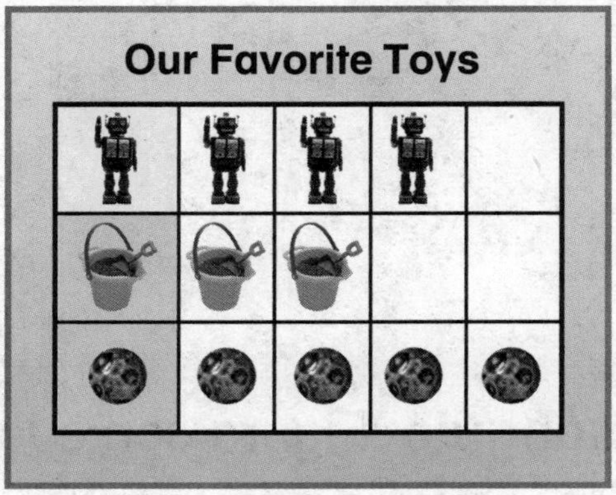

gráfica con imágenes Gráfica que tiene diferentes imágenes para ilustrar la información recopilada.

place value The value given to a digit by its place in a number.

53

5 is in the tens place.
3 is in the ones place.

valor posicional Valor de un *dígito* según el lugar en el número.

53

5 está en el lugar de las decenas.
3 está en el lugar de las unidades.

plus (+) The sign used to show addition.

$4 + 5 = 9$

plus sign

más (+) Símbolo para mostrar la suma.

$4 + 5 = 9$

signo más

Pp

position Tells where an object is.

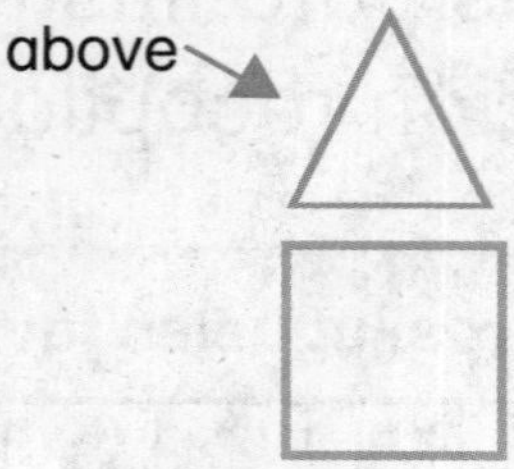

posición Indica dónde está un objeto.

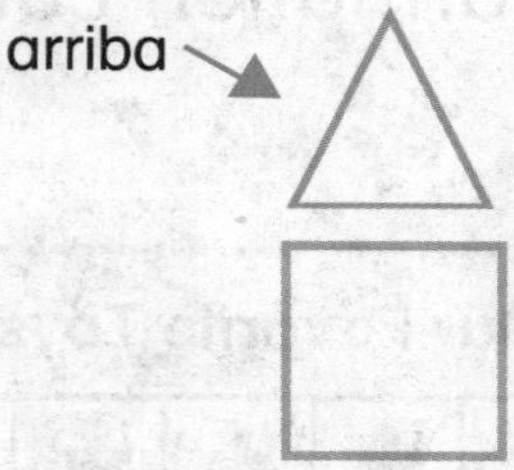

Rr

rectangle A shape with four sides and four corners.

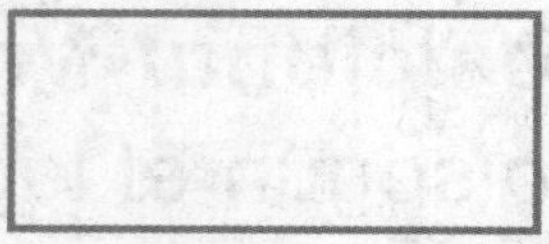

rectángulo Figura con cuatro lados y cuatro esquinas.

rectangular prism A three-dimensional shape with 6 faces that are rectangles.

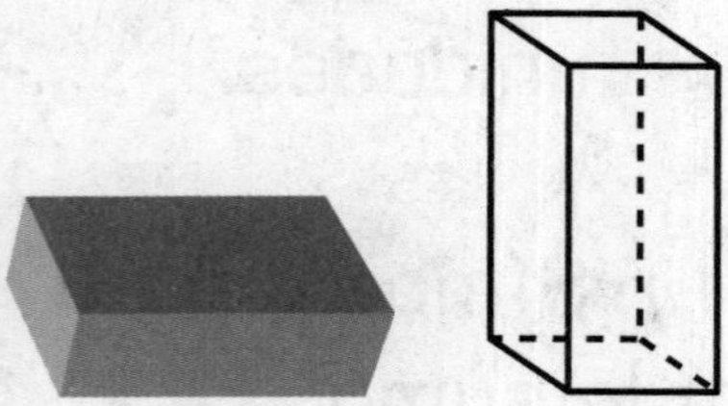

prisma rectangular Una figura tridimensional con 6 caras que son rectángulos.

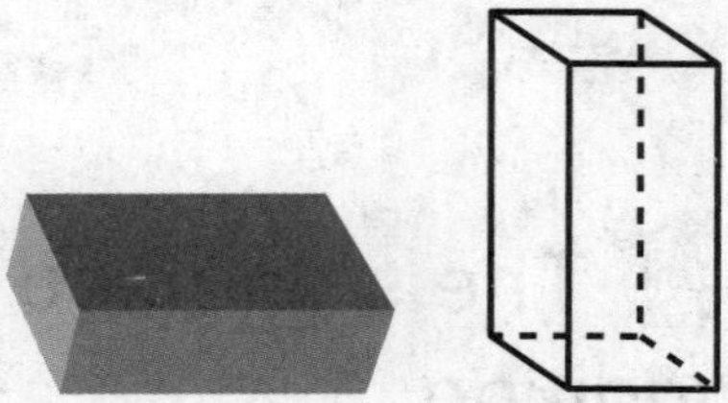

regroup To take apart a number to write it in a new way.

1 ten + 2 ones becomes 12 ones.

reagrupar Separar un número para escribirlo en una nueva forma.

1 decena + 2 unidades se convierten en 12 unidades.

related fact(s) Basic facts using the same numbers. Sometimes called a *fact family*.

$4 + 1 = 5$ $5 - 4 = 1$

$1 + 4 = 5$ $5 - 1 = 4$

operaciones relacionadas Operaciones básicas en las cuales se usan los mismos números. También se llaman *familias de operaciones.*

$4 + 1 = 5$ $5 - 4 = 1$

$1 + 4 = 5$ $5 - 1 = 4$

repeating pattern

patrón repetitivo

Ss

short (shorter, shortest) To compare length or height of two (or more) objects.

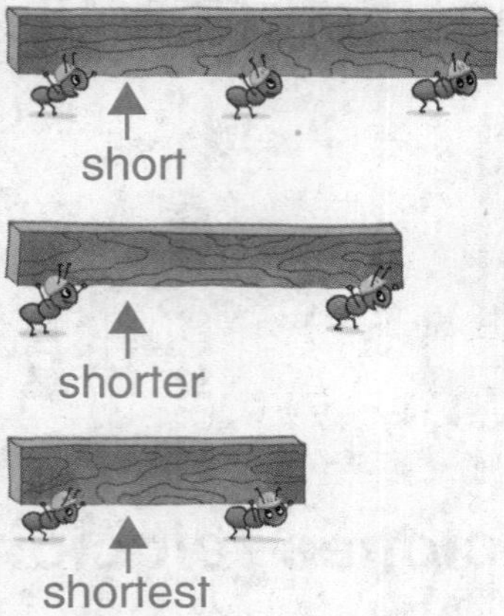

corto (más corto, el más corto) Comparar la longitud o la altura de dos (o más) objetos.

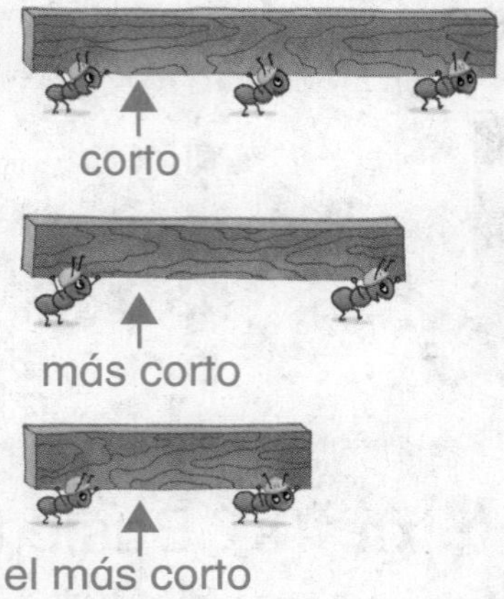

side

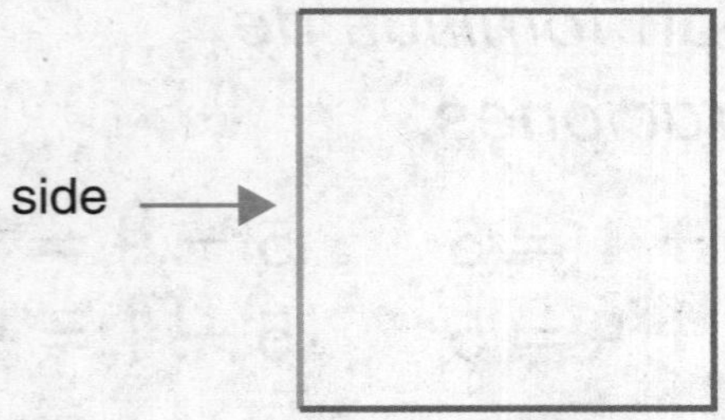

lado

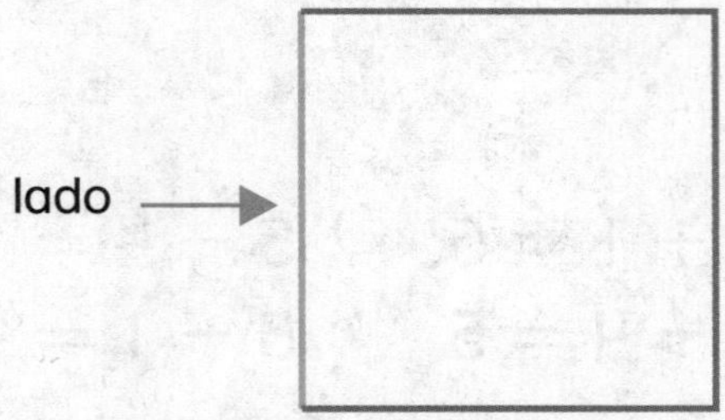

sort To group together like items.

clasificar Agrupar elementos con iguales características.

sphere A solid shape that has the shape of a round ball.

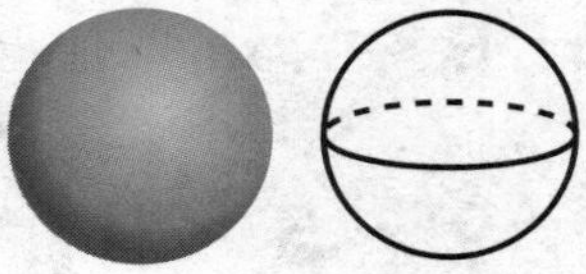

esfera Un sólido con la forma de una pelota redonda.

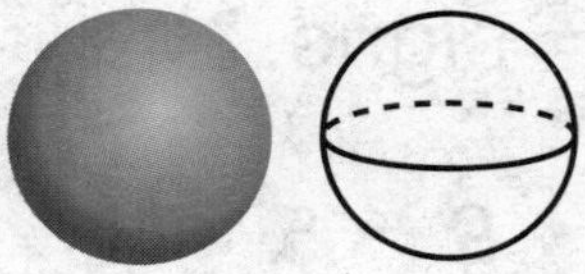

square A rectangle that has four equal sides.

cuadrado Rectángulo que tiene cuatro lados iguales.

subtract (subtracting, subtraction) To take away, take apart, separate, or find the difference between two sets. The opposite of addition.

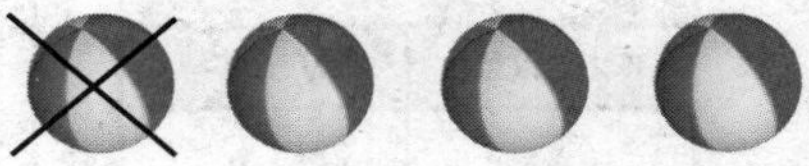

$4 - 1 = 3$

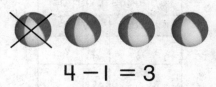

restar (resta, sustracción) Eliminar, quitar, separar o hallar la diferencia entre dos conjuntos. Lo opuesto de la suma.

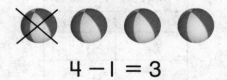

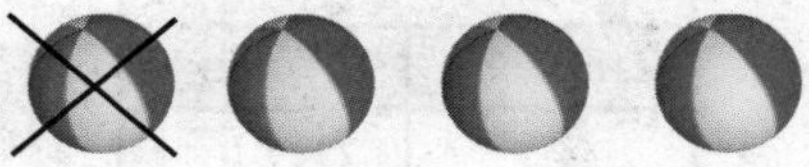

$4 - 1 = 3$

Ss

subtraction number sentence An expression using numbers and the − and = signs.

$$9 - 5 = 4$$

enunciado numérico de resta Expresión en la cual se usan números con los signos − e =.

$$9 - 5 = 4$$

sum The answer to an addition problem.

$$2 + 4 = 6$$

↑ sum

suma Resultado de la operación de sumar.

$$2 + 4 = 6$$

↑ suma

survey To collect data by asking people the same question.

Favorite Foods	
Food	Votes
	卌
	|||
	卌 |||

This survey shows favorite foods.

encuesta Recopilación de datos haciendo las mismas preguntas a un grupo de personas.

Comidas favoritas	
Comida	Votos
	卌
	|||
	卌 |||

Esta encuesta muestra las comidas favoritas.

Tt

tall (taller, tallest)

tall

alto (más alto, el más alto)

alto

tally chart A way to show data collected using tally marks.

Favorite Foods	
Food	Votes
	卌
	|||
	卌 |||

tabla de conteo Forma de mostrar los datos recopilados utilizando marcas de conteo.

Comidas favoritas	
Comida	Votos
	卌
	|||
	卌 |||

tens The numbers in the range 10–99. It is the place value of a number.

53

5 is in the tens place.
3 is in the ones place.

decenas Los números en el rango del 10 al 99. Es el valor posicional de un número.

53

5 está en el lugar de las decenas.
3 está en el lugar de las unidades.

three-dimensional shape
A solid shape.

figura tridimensional
Un sólido.

trapezoid A four-sided plane shape with only two opposite sides that are parallel.

trapecio Figura de cuatro lados con solo dos lados opuestos que son paralelos.

triangle A shape with three sides.

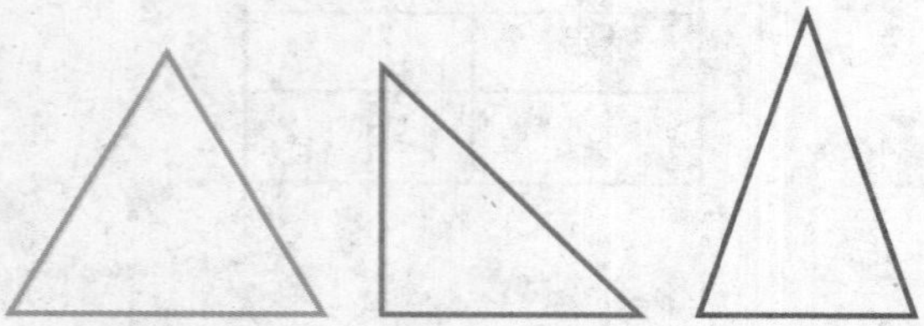

triángulo Figura con tres lados.

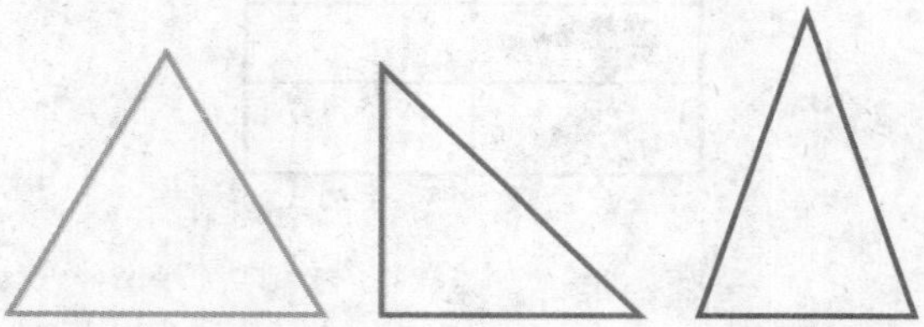

true Something that is a fact. The opposite of false.

verdadero Algo que es cierto. Lo opuesto de falso.

two-dimensional shape The outline of a shape such as a triangle, square, or rectangle.

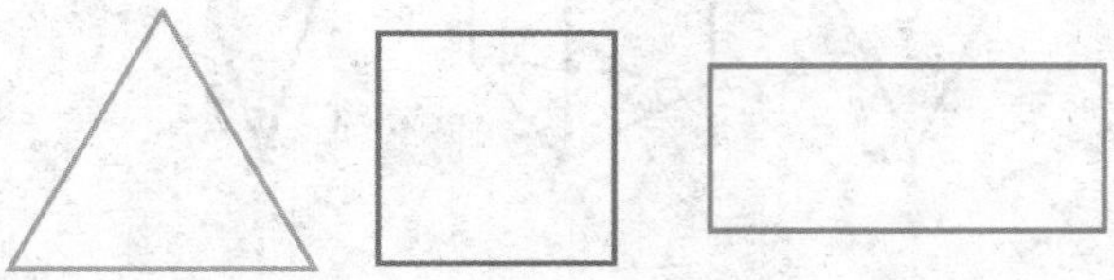

figura bidimensional Contorno de una figura como un triángulo, o un cuadrado rectángulo.

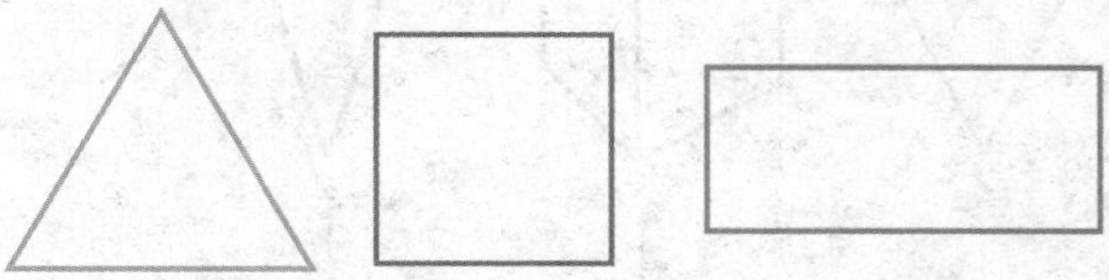

Uu

unit An object used to measure.

unidad Objeto que se usa para medir.

Vv

Venn diagram A drawing that uses circles to sort and show data.

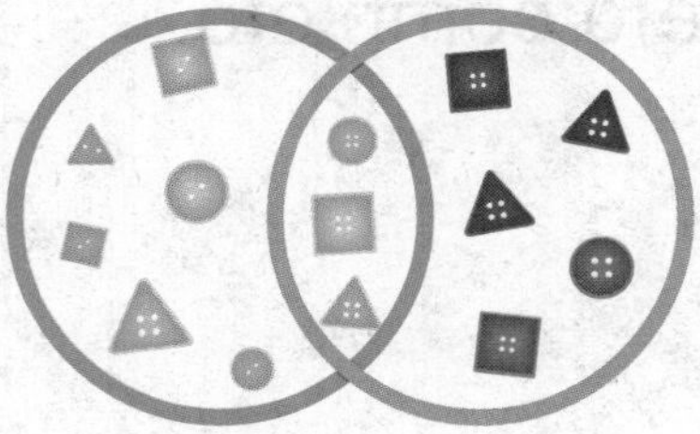

diagrama de Venn Dibujo que tiene círculos para clasificar y mostrar datos.

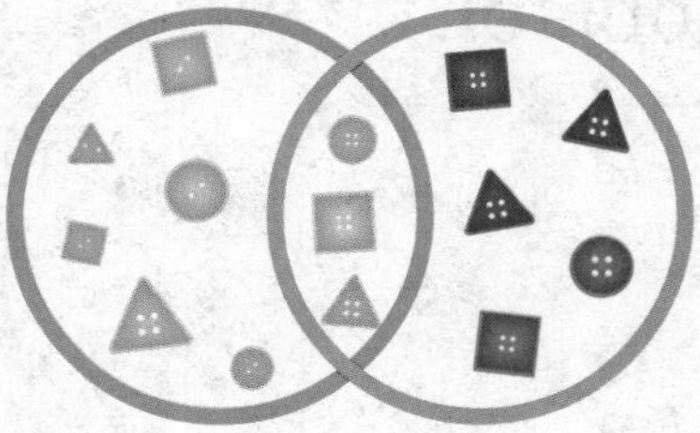

Vv

vertex

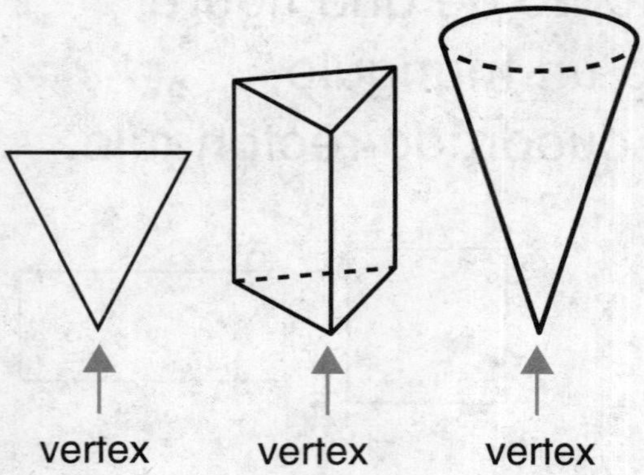

vértice

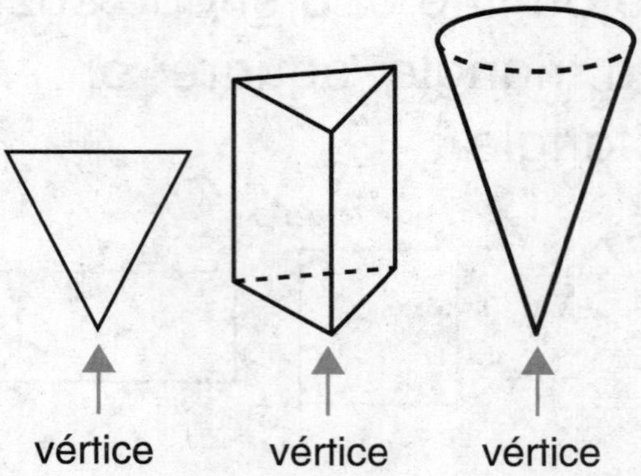

Ww

weight

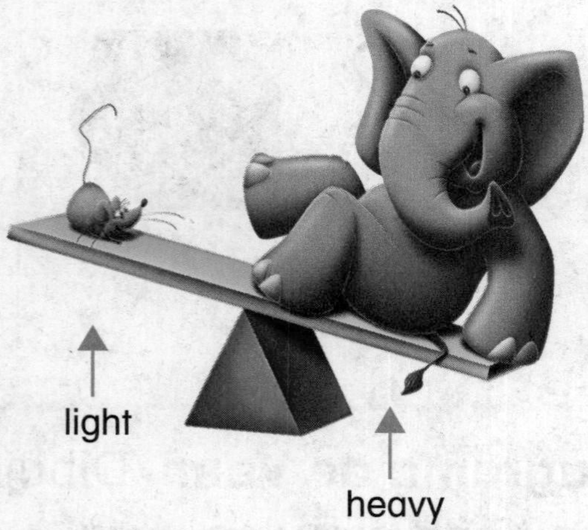

peso

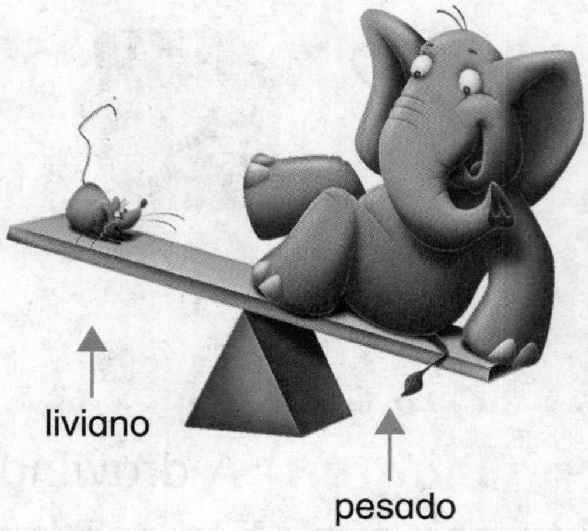

whole The entire amount of an object.

el todo La cantidad total o el objeto completo.

Yy

year

January

S	M	T	W	T	F	S
						1
2	3	4	5	6	7	8
9	10	11	12	13	14	15
16	17	18	19	20	21	22
23	24	25	26	27	28	29
30	31					

February

S	M	T	W	T	F	S
		1	2	3	4	5
6	7	8	9	10	11	12
13	14	15	16	17	18	19
20	21	22	23	24	25	26
27	28					

March

S	M	T	W	T	F	S
		1	2	3	4	5
6	7	8	9	10	11	12
13	14	15	16	17	18	19
20	21	22	23	24	25	26
27	28	29	30	31		

April

S	M	T	W	T	F	S
					1	2
3	4	5	6	7	8	9
10	11	12	13	14	15	16
17	18	19	20	21	22	23
24	25	26	27	28	29	30

May

S	M	T	W	T	F	S
1	2	3	4	5	6	7
8	9	10	11	12	13	14
15	16	17	18	19	20	21
22	23	24	25	26	27	28
29	30	31				

June

S	M	T	W	T	F	S
			1	2	3	4
5	6	7	8	9	10	11
12	13	14	15	16	17	18
19	20	21	22	23	24	25
26	27	28	29	30		

July

S	M	T	W	T	F	S
					1	2
3	4	5	6	7	8	9
10	11	12	13	14	15	16
17	18	19	20	21	22	23
24	25	26	27	28	29	30
31						

August

S	M	T	W	T	F	S
	1	2	3	4	5	6
7	8	9	10	11	12	13
14	15	16	17	18	19	20
21	22	23	24	25	26	27
28	29	30	31			

September

S	M	T	W	T	F	S
				1	2	3
4	5	6	7	8	9	10
11	12	13	14	15	16	17
18	19	20	21	22	23	24
25	26	27	28	29	30	

October

S	M	T	W	T	F	S
						1
2	3	4	5	6	7	8
9	10	11	12	13	14	15
16	17	18	19	20	21	22
23	24	25	26	27	28	29
30	31					

November

S	M	T	W	T	F	S
		1	2	3	4	5
6	7	8	9	10	11	12
13	14	15	16	17	18	19
20	21	22	23	24	25	26
27	28	29	30			

December

S	M	T	W	T	F	S
				1	2	3
4	5	6	7	8	9	10
11	12	13	14	15	16	17
18	19	20	21	22	23	24
25	26	27	28	29	30	31

año

enero

d	l	m	m	j	v	s
						1
2	3	4	5	6	7	8
9	10	11	12	13	14	15
16	17	18	19	20	21	22
23	24	25	26	27	28	29
30	31					

febrero

d	l	m	m	j	v	s
		1	2	3	4	5
6	7	8	9	10	11	12
13	14	15	16	17	18	19
20	21	22	23	24	25	26
27	28					

marzo

d	l	m	m	j	v	s
		1	2	3	4	5
6	7	8	9	10	11	12
13	14	15	16	17	18	19
20	21	22	23	24	25	26
27	28	29	30	31		

abril

d	l	m	m	j	v	s
					1	2
3	4	5	6	7	8	9
10	11	12	13	14	15	16
17	18	19	20	21	22	23
24	25	26	27	28	29	30

mayo

d	l	m	m	j	v	s
1	2	3	4	5	6	7
8	9	10	11	12	13	14
15	16	17	18	19	20	21
22	23	24	25	26	27	28
29	30	31				

junio

d	l	m	m	j	v	s
			1	2	3	4
5	6	7	8	9	10	11
12	13	14	15	16	17	18
19	20	21	22	23	24	25
26	27	28	29	30		

julio

d	l	m	m	j	v	s
					1	2
3	4	5	6	7	8	9
10	11	12	13	14	15	16
17	18	19	20	21	22	23
24	25	26	27	28	29	30
31						

agosto

d	l	m	m	j	v	s
	1	2	3	4	5	6
7	8	9	10	11	12	13
14	15	16	17	18	19	20
21	22	23	24	25	26	27
28	29	30	31			

septiembre

d	l	m	m	j	v	s
				1	2	3
4	5	6	7	8	9	10
11	12	13	14	15	16	17
18	19	20	21	22	23	24
25	26	27	28	29	30	

octubre

d	l	m	m	j	v	s
						1
2	3	4	5	6	7	8
9	10	11	12	13	14	15
16	17	18	19	20	21	22
23	24	25	26	27	28	29
30	31					

noviembre

d	l	m	m	j	v	s
		1	2	3	4	5
6	7	8	9	10	11	12
13	14	15	16	17	18	19
20	21	22	23	24	25	26
27	28	29	30			

diciembre

d	l	m	m	j	v	s
				1	2	3
4	5	6	7	8	9	10
11	12	13	14	15	16	17
18	19	20	21	22	23	24
25	26	27	28	29	30	31

Zz

zero The number zero equals none or nothing.

cero El número cero es igual a nada o ninguno.

Name ....

Work Mat I: Ten-Frame

Work Mat 2: Ten-Frames

Name ....

Work Mat 3: Part-Part-Whole

Part	Part	Whole

Work Mat 4: Number Lines

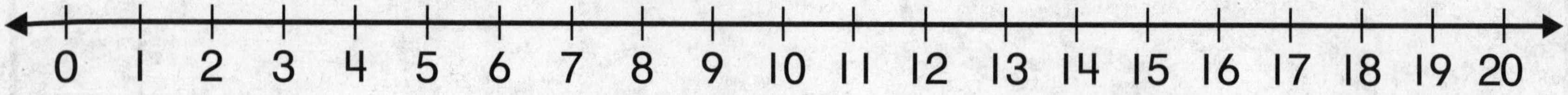

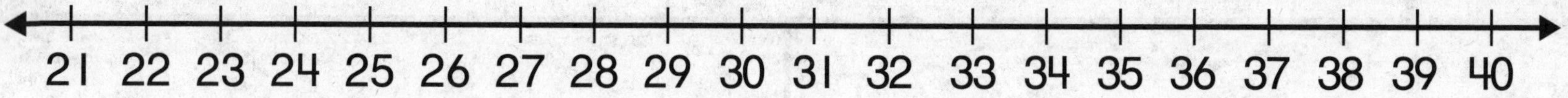

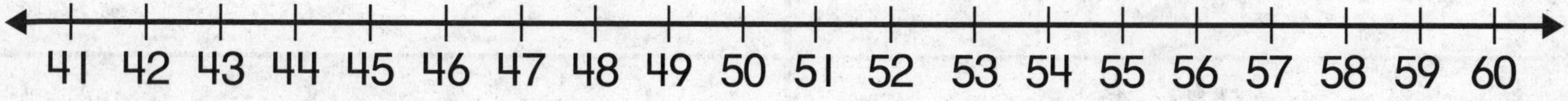

Name ....

Work Mat 5: Number Lines

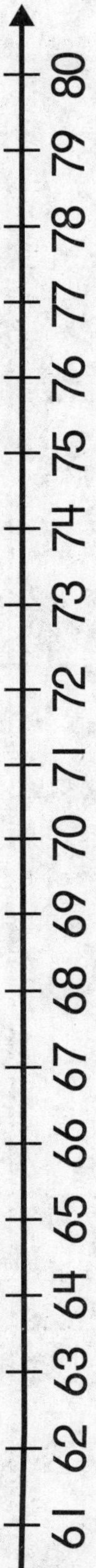

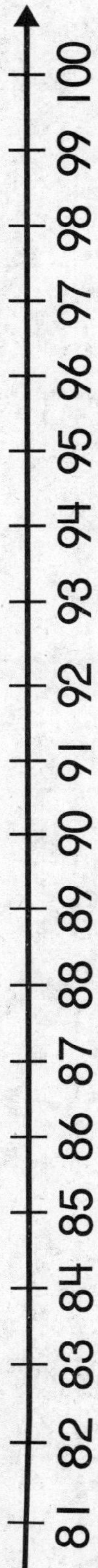

Work Mat 6: Grid

Name

Work Mat 7: Tens and Ones Chart

Tens	Ones

WM7

Work Mat 8: Hundreds, Tens, and Ones Chart

Hundreds	Tens	Ones